RECOLLECTIONS OF A
Pacific
Entomologist
1925–1966

WITH PHOTOGRAPHS BY THE AUTHOR

R.W. Paine

Australian Centre for International Agricultural Research
Canberra 1994

The Australian Centre for International Agricultural
Research (ACIAR) was established in June 1982 by an
Act of the Australian Parliament. Its primary mandate is to
help identify agricultural problems in developing countries
and to commission collaborative research between
Australian and developing country researchers in fields
where Australia has special competence.

Where trade names are used this does not constitute
endorsement of nor discrimination against any product
by the Centre.

ACIAR MONOGRAPH SERIES

This peer-reviewed series contains the results of
original research supported by ACIAR, or material
deemed relevant to ACIAR's research and development
objectives. The series is distributed internationally, with
an emphasis on developing countries.

Paine, R.W. 1994. Recollections of a Pacific Entomologist
1925 - 1966. ACIAR Monograph No 27. 120pp.

ISBN 1 86320 106 8

Technical editing and production:
Arawang Information Bureau Pty Ltd, Canberra
Cover: BPD Graphic Associates, Canberra
in association with Arawang Information Bureau Pty Ltd
Printed by The Craftsman Press Pty Ltd, Burwood, Victoria.
ACIAR acknowledges the generous support of the Paine family
in the compilation of this book.

Foreword

Long before agricultural sustainability entered common parlance, or hazards associated with misuse of pesticides captured headlines, environmentally friendly biological control of introduced pests was being pursued in several countries by a tiny band of amateur and professional naturalists. In Australia, we are all familiar with the biological control of prickly pear by the introduced *Cactoblastis* moth, and of the European rabbit by the introduced myxoma virus. There are a further 50 or so success stories here, and many others elsewhere.

ACIAR, from its formation in 1982, has sought to re-kindle lapsed interest in biological control in the western Pacific and, more recently, to augment the considerable momentum already generated in Southeast Asia. It is noteworthy therefore that, even before successful control was achieved of prickly pear, work based on Fiji in the late 1920s was already at the forefront of world biological control activities. Many of the projects studied by Ron Paine and his colleagues are touched on in his delightful and evocative reminiscences.

Some projects were highly successful, such as the classical eradication of the *Levuana* moth attacking coconut palms, whereas others, such as that on the banana scab moth, have not yet succumbed to biological control. Although the essential scientific facts of the investigations are mostly documented in various publications and reports, the horrendous logistics and other problems of those days are scarcely mentioned. These included considerable personal hardships; absences of means of speedy communication; lack, for long periods, of professional interactions and access to scientific literature; and lengthy sea voyages from Southeast Asia to Fiji to

3

deliver the rapidly dwindling stock of natural enemies. All these and many more were accepted as part of a day's work: far fewer obstacles than them would be daunting to most biological workers today.

Thus 'Recollections of a Pacific Entomologist 1925–1966' provides not only a valuable insight into the historic setting in which the investigations took place, but also biological information that is not available elsewhere and which deserves a place in the published record.

D.F. Waterhouse
Canberra, 1994

Contents

Foreword   3

Preface   6

PART I — 1925–1934   9
  Starting in entomology   10
  Fiji and the 'Levuana Campaign'   16
  Ladybirds for coconut scale control   27
  Australia, Solomon Islands and home leave   30
  The search for coconut spike moth parasites in Java   33
  Parasites of coconut leaf-mining beetle   39
  Introducing *Megarhinus*, a predatory mosquito   44
  Shipment of parasites to Fiji, 1931   48
  Home leave, 1932   54
  Return to Fiji, via Singapore and Java   57

INTERLUDE — 1935–1955
  A market gardener in Scotland   68

PART II — 1956–1966
  Return to the Pacific   74
  The search for banana scab moth parasites in
  New Guinea   78
  Parasites for coconut flat moth   88
  Fiji and home leave, 1958   91
  A new contract — work in Singapore, Malaya, and
  Indonesia   93
  The *Graeffea* Project   101
  The Rhinoceros Beetle Project   106
  A last farewell to the Pacific   119

  Publications: R.W. Paine   120

Preface

This is an account of the travels and experiences of an applied entomologist between 1925 and 1966, mostly in the Indo-Pacific region. I was, for the most part, based in Fiji, and my job was to travel to other parts of the Indo–Pacific region to seek, collect, and send back to Fiji natural enemies of insect pests threatening Fiji's main crops.

My work in these biological control campaigns was spread over two very distinct periods. The first, 1925–34, involved surface travel by ship and motor car. Short flights at low level were taken in Java on three occasions, but all transportation of insect parasites was by ship; sometimes involving different shipping routes to achieve the lengthy journey from Java, or Malaya, to Fiji.

The second period, 1956–66, was after I had forsaken entomology and indulged in market-gardening, during which (including the war years) I had only fringe contact with entomologists, although these were certainly enlivening. In 1956 I re-embarked on Pacific field entomology for three contracts with the Fiji Government (mostly at the U.K. taxpayers' expense) and one for the South Pacific Commission (SPC) terminating in 1966. During these years, transport for personnel was almost entirely by air, as well as, in small parcels, for insect parasites.

The essential differences in the methods used could hardly have been more diverse. In the early period insects had to be *bred* during transit if success was to be reasonably assured. In the later period, when only one or two days travel was needed, the only precaution was to avoid chill in transit. Insect consignments had to be kept in the cabin of aircraft and, if flights were changed in Sydney during the Australian winter, the packages had to be kept in a warm place and

protected from quarantine spraying.

The target pests involved were five leaf-feeding coconut insects, one attacking coconut flowers, and one causing scab on bananas. The last pest to be studied was the rhinoceros beetle, which also causes problems in coconut. As a sideline, I also studied mosquitoes, in particular the potential of predaceous *Megarhinus* species for control of anopheline mosquitoes.

The systematic placing of some of the pests is still, as far as I know, not fully resolved. For facility of contemporary references it seems best to use the nomenclature by which my colleagues and I knew them at the time, as tabulated below.

Species	Insect Order: Family	Common name
Levuana iridescens Baker	Lepidoptera: Zygaenidae	Coconut moth
Aspidiotus destructor Signoret	Hemiptera: Diaspididae	Coconut scale
Promecotheca reicheii Baly	Coleoptera: Chrysomelidae	Coconut leaf-mining beetle
Agonoxena argaula Meyrick	Lepidoptera: Agonoxenidae	Coconut flat moth
Tirathaba trichogramma Meyrick	Lepidoptera: Pyralidae	Coconut spike moth
Nacoleia octasema Meyrick	Lepidoptera: Pyralidae	Banana scab moth
Graeffea crouanii Le Guillou	Phasmatodea: Phasmatidae	Coconut stick insect
Oryctes rhinoceros L.	Coleoptera: Scarabaeidae	Rhinoceros beetle

PART I — 1925–1934

Starting in Entomology

After a boyhood in the depths of the Norfolk countryside the idea of bug-hunting really started when I was at Eton. M.D. 'Piggy' Hill, the inspiring science master there in 1920, thought it a good plan if I had a talk with a former pupil of his who had experience of applied entomology in South Africa.

S.A. Neave, this former pupil, told me the sort of work an applied entomologist would do was to protect food crops from insect pests. As this was mostly in tropical countries, then largely British Colonial Territories, I decided to make this my career. For me it provided a means to explore the world, at least large parts of it. Even at an early age I had an explorer's urge. The popular notion of an entomologist going about waving a butterfly net had no special attraction for me; I would possibly have been a better entomologist if it had.

M.D. Hill was a quite delightful person. It was rather a puzzle how he had got his nickname. There seemed a possible clue when one of his house boys told me that when Piggy ate stewed prunes, he tucked the stones, hamster fashion, inside his upper lip. Then, holding a hand to field any stragglers, he would eject them in a stream onto his plate. Perhaps he had other domestic habits which didn't concur with conventional table manners, but none that I know of.

Another Eton master, Kenneth Fisher, who was not successful in teaching me much chemistry, did teach me how to find wood wrens' nests on the ground at Burnham Beeches. His eldest son, James, became a well known ornithologist. Kenneth the father, let me drive his 'Swift' car. Later he became headmaster of Oundle. He was one of Peter Scott's mentors and a charming individual.

By sea and rail to the South Pacific

After three years at Cambridge for a degree in Natural Science, I had barely embarked on a course of post-graduate entomology in the autumn of 1924, when an entomologist from Canada, J.D. Tothill, visited Cambridge University to enlist two graduate assistants for a campaign to control a pest of coconuts in Fiji. Frank Balfour-Brown, who ran the course, urged me to apply.

It was an exciting day for me. I rushed home to tell my parents and to get out the *Times Atlas* to see just where in the Pacific Ocean the Fiji Islands were. I hadn't to wait many days before an official letter arrived offering me a temporary appointment as an assistant entomologist in Fiji for a period of eighteen months to two years. Included were official reservations. If I was found unsuitable the Governor could cancel my appointment at any time, with my return passage to England; but only if I hadn't misbehaved.

I was too stimulated by the prospects of this job to do more than skim through all the regulations and conditions in the letter. It concluded by asking if I was prepared to proceed to the colony by a steamer leaving Liverpool on 16th January 1925. It bore the signature 'H.F. Batterbee'.

The remainder of that term was spent mostly away from Cambridge, kitting myself with the sort of clothing recommended for the tropics and the luggage suitable to contain it. I even took a shot gun and golf clubs, as there seemed little restriction on weight or bulk of baggage for a journey mostly by ship.

I sailed by the *Montcalm* from Liverpool. As things turned out it was inappropriately named

Trinity College, Cambridge

for this winter Atlantic crossing. On board I met for the first time T.H.C. Taylor, Tothill's other assistant, a graduate of Reading University. We hit it off from the start and thereafter developed a close lifelong friendship.

The ship suffered some damage from a mid-Atlantic storm. For those affected with seasickness it must have been a miserable two days. As I am fortunately spared this affliction, I couldn't help being amused by the attempts of passengers to play cards in the lounge. Not only did the cards make frequent journeys across the floor, but a rather bulky Presbyterian minister had a chair leg break under him. After a steward had picked him up on to another, that gave way as well. The bridge players then gave up and sought safer places for themselves.

A very pasty-faced man, sprawling over a sofa, made a gurgling noise and then remarked: 'That sounds dangerous, doesn't it?' The noise was so loud everyone in the lounge heard it and a prim-looking lady looked so horrified I could contain my mirth no longer and retreated, to simmer outside the room.

After the storm was over I met an attractive young lady called the Hon. Evelyn Gardiner. In a casual conversation I discovered she knew one of my closest school friends, Ian Burn. Her mother, Lady Beauclerk, was on the passenger list, but Miss Gardiner was seldom with her, and usually in the company of a tall military, but rather debauched-looking, male friend. But that didn't deter her from inviting me for a drink with them in her cabin. I began to wonder what Taylor might think about me consorting with such rather high-flown folk. He had been seen off at Liverpool by a girl, to whom I learned later he was engaged, so he had some defence against female

Across the Rockies by rail

company on the voyage. I had none, and being an increasing distance from any parental restraint, fell victim to my susceptibilities.

After we landed at St John, New Brunswick, Evelyn got on the train to Ottawa, so I had to say goodbye to her. She and her mother were to be guests of the Governor General.

Taylor and I got on the transcontinental train for Vancouver. It wasn't until I had watched a TV program in 1988 about Evelyn Waugh that I discovered that he had for a short time been married to this same Evelyn Gardiner.

The four-and-a-half days train journey was delightful — all unknown and exciting. At Montreal, where it was −7°F, it felt devilishly cold — a damp cold, much more chilling than the −30°F dry cold at Winnipeg. But there I was foolish enough to touch the metal railway line which blistered my hand with its frozen chill. I developed a taste for hot cakes and maple syrup for breakfast, and clam chowder for lunch. There was a small observation car at the rear of the train in which I sat watching the rails coming together in the far distance across the prairie. As we glided through the Rockies I tried taking snaps of high peaks, snow-clad and majestic. The way the track snaked its way from the summit down to the Fraser canyon kept me on tenterhooks, looking out first one side then the other.

I had an uncle and family living at Alberni on Vancouver Island. He met the train at Vancouver and after a night there he took me to spend a few days at his home. There I got the first batch of mail from home.

My father enclosed one from my Eton housemaster, 'Jelly' Churchill. Like Piggy Hill, E.L. Churchill got what seemed a most inappropriate nickname. He said the Colonial Office had asked him what he thought of me. He told them: 'He always wanted to do what I told him not to do — a confirmed breaker of rules. It occurred to me that this characteristic renders him eminently suitable for dealing with savages. I shall always believe that my testimony got him the job.'

The *Aorangi* of the Canadian–Australian Royal Mail Line sailed from Southampton via Panama to start her maiden voyage from Vancouver on 6th February. Before leaving the U.K. her intended stewards had gone on strike and a lot of inexperienced men were hurriedly signed on for that work. Passengers had been ill-served and were disgruntled by the time the ship reached Vancouver. She was in port there for several days. Taylor joined her there, and I a few days later at Victoria. Had she sailed a week later, on the 13th, superstitious folk would have been justified in thinking it the cause of the exceptional storm she struck on her way to Honolulu. However, it was not that date; but the storm disabled the wireless and damaged part of the deck. At intervals the sun shone and I got some realistic snaps of tall ocean waves. On Sunday morning the church service was cancelled as it was too rough for the captain to leave the bridge. A plate glass shield on the upper deck was broken by huge waves. It wasn't my preconceived idea of the Pacific Ocean, but I rather enjoyed it.

As we got nearer the Hawaiian Islands the sea smoothed and the moon shone bright for our last night at sea. My excitement was tinged with apprehension. My new work was to begin. I wondered how I would cope with it.

The high seas, from the deck of the Aorangi

An introduction to Hawaii

We were met at Honolulu by Fred Muir, chief entomologist of the Hawaiian Sugar Planters Association (HSPA). I had heard him lecture at Cambridge, so his trim, greyish appearance was familiar. He drove the two of us to the Macdonald Hotel, a small unpretentious building charging $60 a month (£14 at $4.30 as it was then). It was mid-February and we found the relentless 80°F a bit trying. We sweated profusely and bought frequent ice drinks at 'drug stores' and travelled in 'street cars'.

Dr F.X. Williams, a quiet, pleasant-mannered, unmarried entomologist, was the only HSPA man staying at the hotel. The other guests were American small business folk and two Scots working in banks. We got to know them after a picnic with them during the first weekend.

In 1925 one walked through a stretch of flat wasteland behind Honolulu town before reaching the foothills of the Tantalus Ranges. Taylor and I spent hours collecting from wild *Amaranthus* plants small green caterpillars which were attacked by an ichneumonid. But our main source of material was the coconut leaf-roller, *Omiodes blackburni*. From its caterpillars we bred parasites: a *Chelonus* and a tachinid. We rested sitting on the dry, Lantana-clad foothills. I now grow a potted plant of *Lantana camara* and the smell of the flowers revives a memory of Honolulu in February 1925.

We did several hikes in the Tantalus hills with Williams and O.H. Swezey. Swezey picked and ate guava fruits growing in the bush. When, inadvertently, he ate one infested with fruit-fly larvae it made him sick.

During our last weekend in Hawaii, Muir, Swezey and Williams took us on a long hike to Mt Kaala, Oahu's highest summit. I took a photo of the three with Taylor, near the top. In the eighties it became a sought-after photograph, since published in 'Antenna'. From that 4000 ft situation we could get a good view of the naval base at Pearl Harbour, at that time such a peaceful prospect.

Fred Muir was a good host to us: he took us to Waikiki beach for surfing; drove us to that dramatic cliff top, Nuuanu Pali, and had us to tea in his garden. Fred's wife was a daughter of Dr David Sharp, author of one insect volume of the 'Cambridge Natural History'. Fred Muir impressed on us the importance of taking entomology with the utmost devotion, not just to dabble in it. Taylor certainly fulfilled this injunction. I'm not too sure that I did; for after marriage, we didn't want to live in the tropics, so gave it up.

After our month in Honolulu, Taylor and I sailed for Fiji with an assortment of parasitic wasps and flies, not to mention all the scented garlands (leis) draped round the necks of departing visitors. We were given a large box for our insects which had to be kept in our cabin. The voyage on the *Makura* took eight days. I'm glad to say there was no 'Father Neptune' tomfoolery when we crossed the equator and for our initial experience a day was missed as the date line was crossed just before reaching Fiji.

Fiji and the 'Levuana Campaign'

When we stepped on to the Suva wharf on a dull, damp morning we were met by A.B. Ackland. 'Hop', as he was known, was Secretary of the Department of Agriculture. He drove us to our headquarters in a small bungalow near Government House. It had been built for the Prince of Wales's visit the previous year, but he never used it, preferring to sleep on his ship. It suited us well enough as an office, but had no laboratory facilities other than a wash basin.

Tothill was indisposed the day we arrived, but on the next day he was better and took us to his house. There we met Mrs Tothill, a quiet-voiced American, a trained botanist and obviously very interested in Fiji's plants. The eldest of their three children, Jessie, was seven. There were two sons, Tom and Allan.

The idea of bringing the parasites from Hawaii was more to give us practice in handling insects on a ship than for the control of the coconut moth, *Levuana iridescens*. They were duly released but few, if any, would have become established.

It was assumed, at that time, that *Levuana* had been introduced to Fiji. It had no parasites there of any significance and had flared up and spread since 1920, from Viti Levu, the main island,

A fine stand of coconut palms

Agriculture Department building, Fiji

eastwards to threaten the main copra-producing islands. Copra, after sugar, was Fiji's most valuable export.

The 'Levuana Campaign', as it was called, seemed to entail a search for the original home of the pest and the hope that effective parasites would be keeping it under control there. To give effect to this plan, Tothill decided that one of his assistants should undertake the search and the other should remain in Fiji to study the pest. This was something which should be decided by the toss of a coin.

It was going to be one of those moments in life when the decision could have a profound influence on one's future. Both Taylor and I were keen to undertake the overseas search. Tothill, I think, must have sensed our feelings, and to dramatise the matter still more he said it would be decided by the best of three tosses. We held our breath. I won the first; 'butterflies' flew hopefully within me. Taylor won the second, they flew even more indecisively. Then Taylor won the final toss ... so that was it. I felt keenly disappointed, but in hindsight I think the right decision had been made. Taylor was to do a splendid job away from Fiji, resulting in his bringing by devious routes from Malaya the parasite which put paid to the *Levuana* moth. For me there was plenty of interest in the local study of both the pest itself and plenty else as the months spun past. Also, there was the advantage of having Tothill to work with. He was a splendidly informal boss, and good company on such trips as we did together.

Shortly after that had been decided, the Coconut

Committee, made up of Sir Maynard Hedstrom, Edward Duncan and Tothill, went, with Taylor and me, to Moturiki, one of the recently infested islands. We travelled in style on the Governor's ship *Pioneer*. Hedstrom was one of the so-called 'Big Four', owning a department store, Morris, Hedstrom Ltd, which handled much of the copra from holdings of the indigenous Fijians, as well as from the big four's own estates.

We had to experience one of the customary native welcoming ceremonies and I felt for the first time the discomfort of sitting cross-legged on the floor for more than five minutes. I never acquired this apparently simple skill, I hadn't the right muscles, and on many future occasions insisted on having a chair. Hedstrom spent most of the day fishing. Duncan stayed in the village. The rest of us walked about the island, looking at the much-defoliated coconut palms.

A coconut palm attacked by Levuana

Pioneer, *the Governor's ship, anchored at Nabaratu*

Insectary built for the 'Levuana Campaign'

On returning from Moturiki we put in at a very small, circular island, Leleuvia. The ship's dinghy was unable to reach the shore, so we were carried ashore on the shoulders of Fijians, who with their large feet and spreading toes, seemed unaffected by the rough surface of coral. Leleuvia had also suffered from the recent outbreak of the coconut moth. Between the palms there was a thicket of overgrown shrubs among which were the small hanging nests of the Fiji hornet (*Polistes hebraeus*), that widespread and painful deterrent to field work between late March and early May. Some misguided person brought this to Fiji in the hope it would kill flies.

When we returned to Suva, Taylor set off for New Guinea. I was provided with basic laboratory facilities near an insectary built by the Public Works Department to Tothill's specification. Our project was attached to the Agricultural Department which was centred two miles away near the Suva wharfs.

Tothill had enlisted a young English speaking Fijian, Jakopi (Jacob), to be our factotum — a role he seemed only occasionally to fulfil. He was my constant dogsbody for the next two years. Although sometimes delinquent, he had a subtle intellect. One morning I asked him to get me fifty *Levuana* caterpillars and to be back by two-thirty. He hadn't returned by four, when I had to leave the laboratory and lock up. The next morning I found a cigarette tin crammed with dead caterpillars. In a fit of surprising honesty he said 'Mr Spain (it took weeks for him to drop the 'S') I took one leaf of your tobacco, Sir, only one leaf'. He often offered me cigarettes. This was perhaps my first lesson in the Fiji custom of 'keri keri': ask and you will get. I had a lot to learn, and I fear poor Jacob had none too good a boss for a start. He came to be well versed in Tothill's and my proceedings when studying *Levuana* in the field and considered himself a matai (expert) when telling his friends about the work.

After two months based in Suva studying our pest in the 'wet zone' of Viti Levu, I moved north to Ra Province on the drier side of the island, to a plantation called 'Nanuku' belonging to Charlie Burness.

Charlie lived with his elderly mother, who claimed descent from Robbie Burns. Charlie, born in Fiji, employed Indian and Fijian labour to cut copra and cook and tend to household chores. He was fluent in Hindi as well as Fijian. He was a kind man who didn't fuss and was happy to give me a bed, meals and space to accommodate any insects I wanted to study. He lived in a singlet and shorts which usually looked in need of a washtub.

So extensive was the *Levuana* damage that moths had been ovipositing on banana plants on which the larvae seemed to feed quite happily. It was an ideal situation to study the pest.

Although so abundant at that time in Fiji, the BM [British

'Bay of islands', Nabaratu

Landing cages of B. remota *at Caboni*

Museum (Natural History)] had very few specimens of *Levuana*. I mounted and sent several, but they were still poorly represented, rather unfortunate in view of the later extinction of this species.

During July, Taylor, who had found a number of related zygaenids on zingiberaceous plants in the New Guinea coastal zones, was sent to join H.W. Simmonds in Malaya, where an outbreak of the local coconut leaf moth, *Artona*

catoxantha, was in progress at Butu Gajah. A full account of the collection and shipment to Fiji of larvae of this moth, containing the tachinid fly *Bessa (Ptychomyia) remota*, is given by Tothill in 'The Coconut Moth in Fiji' (Imperial Bureau of Entomology, London, 1930).

Taylor, after successfully getting the tachinid to Fiji, was sent back to New Guinea to seek other parasites in case *B. remota* should need

additional support. Simmonds was sent to Java. *B. remota* was doing so well and spreading so quickly owing to the abundance of its host, I think Tothill was feeling very happy; also preoccupied — having been made Director of Agriculture — and felt the entomologists should stay put for the time being.

I had been given a room and a laboratory in the main agriculture building near the wharf and received an interesting visitor at the end of August. It was Allan P. Dodd, returning to Australia from the southern U.S. with eggs of the *Cactoblastis* moth, which did so much to control prickly pear.

My next visitor was Patrick A. Buxton, returning from the New Hebrides (Vanuatu) to Western Samoa. He was then due to return to the U.K. after completing a survey of medical insects in Samoa. It was certainly Buxton who gave me the impetus to study mosquitoes in Fiji, which was greatly helped by his report illustrating the key structures by which larvae of the dozen or so South Pacific species could be distinguished. I found the collection and study of mosquitoes a fascinating sideline. After collaboration with F.W. Edwards of the BM during leave in 1928, I published 'An Introduction to the Mosquitoes of Fiji'. After the outbreak of war, with American troops stationed in Fiji, it became much in demand. A local insurance agent, called Amos, made a further study of mosquitoes, actually describing a new one, in a revised edition of my paper.

To return to 1925; I had rooms at a boarding house when there was an outbreak of diphtheria in Suva. We were all quarantined, but as I had by then acquired a motorcycle, sent from the U.K., I was allowed to go out on it to collect in the bush away from town. I brought back in my jacket pocket a large millipede, 8.5 inches long, which I produced as the doctor was taking a swab from my throat. It was a rather foolish prank really, as he recoiled violently, thinking it might be venomous. They do, in fact, eject a dark-brown, slightly irritant fluid if roughly handled. I'm glad to say Dr Thompson became a good friend later and we had a common enthusiasm for hill climbing.

Simmonds returned from Java with complaints about the ill treatment he had received in Java. I think he had some justification. Once Tothill had gone to Uganda, the complaints ceased and Simmonds assumed a more cheerful note and became positively friendly. He had built a house in Suva and was a skilful gardener. He also had a wife who sang romantic ballads, accompanying herself on the piano. I was invited to musical evenings to hear 'Rachel Dear' perform.

'Hubert Dear' would then be asked to produce the meringues which he cooked skilfully. Thus the evenings would pass, with never a trace of any stimulant other than a cup of weak coffee.

Much later Simmonds published (privately) a small book of reminiscences, 'My Weapons Have Wings'. It was an odd book with more anecdotes than entomology and oddly with few words per page. In old age he was still in part-time employment by the government, as he was convinced a Madagascan scoliid wasp would control rhinoceros beetle, so continued breeding them until he was well over eighty. It seems quite likely that 'Woggy' Simmonds may figure in Fiji's history as a pioneering entomologist for longer than those brief interlopers Tothill, Taylor and Paine who brought about the control of Fiji's potentially worst coconut pest.

'Levuana Cottage'

During 1926 another young Cambridge graduate came to Fiji as a cadet in the administrative service. Charles Harley Nott and I became close friends. We had both read geology, and joining in bush walks with us was Harry S. Ladd, a young graduate in geology from the U.S. making a study of the geology of Viti Levu.

Nott and I soon tired of life in Suva and decided to rent a small shack at what was then the completely rural Suva Point. This we crudely furnished and lived there for several weeks, going to work on our motorbikes. We had Ladd to a meal one evening which consisted mainly of what he termed a 'six inch stoo'. This was followed by an edition of synthetic creme-de-menthe concocted by Colin Southall (the Government Chemist), from ethyl alcohol with peppermint flavouring. Harry mopped it up.

In spite of this early assault on his stomach, Ladd in later life became a specialist in Washington, D.C. on fossil coral molluscs, and I kept up with him until he died.

After our sojourn at Suva Point, Charles was posted to a country district. An earlier Government House had been burnt down and as its replacement was not yet finished the Governor, Sir Eyre Hudson, and his lady, allowed me to occupy the quarters destined for their second cook. This was an iron roofed shack at an elevated situation on the Suva soapstone plateau. It was surrounded by sensitive grass which tore at my ankles when I walked round the building to the shower. Lady Hudson lent me furniture. I loved it and called it 'Levuana Cottage'.

B. remota had suppressed *Levuana* to such an extent that to keep a stock of the latter for further study we had to enclose a small palm in a cage to keep out the fly. It was all very encouraging.

On a visit to Ovalau to check on the fly's progress there, I met a local resident who for many years had been studying the annual risings of the 'bulolo' reef worm (*Eunice viridis*). It was early in November and I was told that the rising would coincide with the third quarter of the moon. The locally abundant worms which inhabit the coral in certain places, release most of their bodies, containing the male elements, which rise to the surface of the sea in the early morning. I was taken by some locals to the scene of the rising just before dawn. There were many Fijians prancing about near the shore each with a square of fine netting attached to two sticks. Then the rising began, a few worms at first and then a seething mass of their rising bodies. As dawn broke it was an animated scene, the Fijian women shouting and singing as they scooped up the much prized bulolo to eat raw or cooked. I tasted some just as it was caught. To say it tasted like whitebait was to flatter it. It tasted salty but little else. I was glad to have had the chance to witness such an interesting zoological phenomenon.

From Ovalau I was given instructions to go to Koro, a larger island lying further east. Here there was an outbreak of the coconut leaf-miner, *Promecotheca*, which was eventually so well controlled by Taylor's introduction of a eulophid parasite from Java. Back in Suva I met for the first time Ratu Sukuna, a high ranking Fijian, an Oxford graduate who fought in the French Foreign Legion in the 1914–18 war. He came to Levuana Cottage to play bridge. I found him a delightful person. Later he played a leading part in the Department of Native Affairs, was knighted, and after he died, had many memorials named after him.

Over Christmas 1926, Charles Nott and I, with Miss Ida May Smith from the Secretariat Office, set out on a walk to explore the Mendrausuthu mountain range located 30-odd miles inland from the Rewa delta. We had a very hot and exhausting walk, with many river fordings. After a particularly hard scramble up a steep bank we were seated at the top when a hornet stung Charles on the septum of his nose. It must have been extraordinarily painful, giving him every excuse for use of language more colourful than that normally encountered in mixed company.

After a night with the European manager of a banana plantation we climbed over a pass to Naitauvoli, an elevated village blessedly devoid of mosquitoes. We spent a comfortable night in Fiji bures. We were fairly near the range intended as our target, but the Fijians told us about the Waiga River gorge, a spectacular cleft down which they said each of us could be poled on a narrow bamboo raft. It seemed well worth doing. There was plenty of tall bamboo growing near the village and the rafts were soon lashed together.

The river's channel, only ten feet wide in places, had cut through a sandstone hill leaving cliffs 50–60 feet tall on either side. It took about half an hour to get through the gorge and one felt utterly dependent on the skill and ability of the Fijian poling one's raft. It was near the middle of the day when we emerged from the gorge and shortly afterwards heavy rain began to fall and the Waiga became a torrent, making the gorge quite unnavigable. We returned to Suva with the feeling that we had seen something few Europeans could have seen. No-one we spoke to in Suva seemed to know about it.

Early in 1927 another young Cambridge graduate arrived in Suva as a cadet for the Western Pacific High Commission. The Governor of Fiji was also High Commissioner for the Western Pacific, so cadets started their training in Suva.

Ronald Garvey, that latest arrival, Nott and I formed a very amateurish busking troupe, styled 'The Cantab Trio'. We sang popular Fijian songs and some of the well known ballroom songs of the time, accompanying ourselves on ukuleles. Nott and I never knew, nor did Garvey tell us, that he had reached vocal heights as a Lincoln choirboy in a parish where his father was vicar, or that he had

Emerging from the Waiga River rapids

earned a music scholarship at Trent College. These things we learnt only from Garvey's autobiography, 'Gentleman Pauper' (New Horizon (Tanseuros Ltd), Bognor Regis, 1983). However, as a singer he had little respect for such a lowly instrument as a ukulele and took the matter into his own hands by using it as a carpet beater against a length of Tapa in my cottage. Having successfully bust up his uke he gave his mind to more worthwhile projects.

After a continuously upward career in the Colonial Service, Garvey ended it by returning to Fiji as Governor.

In February 1927, the Duke and Duchess of York steamed into Suva in HMS *Renown*. They were en route to Canberra to open the new Parliament House there. New Government House in Suva was not quite built, so a reception for the royal couple was held in the Grand Pacific Hotel to which Government servants were invited. There was a dinner dance. Garvey and I shared a table not far from where the Duchess was seated next to Mr Rushton, the Colonial Treasurer. Ronald said to me across the table, 'I'm going to ask the Duchess for a dance, I don't see why that old so-and-so should monopolise her beauty.' I could hardly believe it. A young cadet, only three weeks in Fiji, having the brass to ask the Duchess for a dance. 'Protocol be damned', thought Garvey. 'The charming Duchess isn't here every day.' So he got up in his tailcoat, walked over and with a slight bow asked the young lady to dance. Elizabeth, Duchess, Queen and finally Queen Mum, smiled graciously at him, then turned to Mr Rushton, who looked daggers at Garvey, and asked him not to reprimand the young cadet (at least that's what we thought she said, as no further action was taken).

The 'Cantab Trio' (RWP in middle) after a fishing trip

Ladybirds for coconut scale control

While all this was going on, Taylor had returned from Java on a ship of Indian coolies infected with cholera, so was quarantined with his collection of ladybirds for control of coconut scale. Before he could hand over the rearing of these to me and proceed on long leave, he had over four weeks quarantine. These coccinellids, a species of *Scymnus*, were not effective for control of coconut scale, *Aspidiotus destructor*; but it was lucky for me that this was not known at the time, because Tothill had been offered accommodation on the *Pioneer* for the occasional trip she did to service the lighthouses. Mrs Tothill and one of their boys came and Tothill thought I should come and take the ladybirds to some of the eastern islands.

It was certainly one of the most delightful trips I had in Fiji, as we visited so many islands. On an uninhabited island in the far north of the group we were able to see *Birgus latro*, the large coconut robber crab, at work tearing open a coconut. As we steamed southwards through the Lau islands we came to Fulanga, an atoll, in the lagoon of which were numerous raised coral islets known as 'Nigger Heads', on which grew the native fan-leafed palm, *Pritchardia pacifica*. From Fulanga we visited Totoya. This seemed to me the most attractive island in the whole of Fiji. It is a horseshoe-shaped island, with the opening to the south, enclosing a large, deep bay. The land rises in low hills, partly tree-clad, from which one can view one or two trim villages with thatched houses set along the coastline. The neighbouring islands, Moala and Matuku, are larger and more hilly, but Totoya is the real pearl of that cluster. I suppose in time it will sprout a tall TV mast and the houses will be tin roofed and have aerials. I'm glad I saw and took photographs of it as it was in 1927.

A village on Totoya Island

On another shorter trip I went on the *Lady Escott* round the north of Viti Levu to visit the Yasawa Islands extending in a chain northwards from Lautoka. One of these was said to be R.M. Ballantyne's *Coral Island*.

There had been a request by the manager of a Unilever plantation for Tothill to visit Solomon Islands and investigate a nutfall problem affecting many coconut estates. He said I should go with him. But there were postponements. Taylor had been married during his leave and was sent to Trinidad to collect and bring to Fiji the coccinellid, *Cryptognatha nodiceps*, which was to successfully control *Aspidiotus* scale. I was to help with the breeding and distribution of these beetles so the Taylors could proceed to Java to study parasites for the coconut leaf-miner.

One of the early releases of this ladybird was at Wainunu, a small plantation with tea as well as coconuts, belonging to a 70-year-old ex-merchant navy captain called Robbie. He was reputed to consume two bottles of Johnny Walker a day. He told me in his broad Scots accent always to take poison well diluted. He told a story of a pussyfoot preacher who had before his audience two glasses, one with water and the other with whisky. He took two live worms, one of which he put into each fluid. The worm in the whisky soon died; the other continued to wriggle about. Then a voice from the audience shouted 'I'm glad you showed me that. I've been bothered by worms for quite some time. Now I know how to get rid of them.'

Over the New Year holiday I had an unfortunate accident. I was leading a party of three across the bay from Suva along a trail in the bush which I knew quite well. We had to leave Suva by launch westwards and spend a night in a village before starting our walk. It was a most uncomfortable night. Rats ran over us as we tried to snatch some sleep. We got an early start along the track in the morning. All went well until my right foot slipped on a damp tree root and I fell on the sharp edge of my bush knife which cut across the fingers of my right hand. After staunching the blood as best we could there was nothing for it but to turn back. It was nine hours before I got to Suva hospital. Dr Harper, the surgeon, did his best to join the tendons of three of my fingers, but it was unsuccessful. Luckily my index finger and thumb were undamaged. I was three weeks in hospital and wrote a rather crudely written letter home with my left hand. Tothill kindly wrote to my parents explaining what had happened. There was no infection and I felt well enough in myself. But for the rest of my life I have had an awkward bent third finger.

After a few weeks I found I could still grip a golf club. I was at the golf club with Charles Nott early in June 1928 when we spotted Kingsford Smith's *Southern Cross* monoplane on his historic trans-Pacific flight from Honolulu. We hurried down into Suva in time to see the plane land in Albert Park. I took photos of the aircraft, one of which the *Times* published. This flight caused a great stir in Fiji. The Fijians sang songs about the 'Waqa Vuka' (flying ship). It seemed suddenly as if Fiji had become much nearer America and Australia.

Charles Kingsford Smith's Southern Cross *about to leave Suva in June 1928*

Australia, Solomon Islands, and home leave

Tothill's and my visit to the Solomons started in August 1928. There had recently been two murders on Malaita: District Officer Bell and his Assistant, Lillies. The enquiry into this was beginning when we reached that island on our survey. On our way to the Solomons we visited Sydney, my first of many later visits to Australia. We had a week there before the *Mataram* sailed for the Solomons. Tothill went off to Melbourne to make various arrangements. I played golf and stayed at the luxurious Rose Bay Golf Club. I also met Dr A.J. Nicholson, entomologist, who gave me some useful tips for preserving insects and mounting caterpillars in life-like condition. Others I met were George Julius, who invented the totalisator, and Ian Clunies-Ross, a veterinarian, who later became one of Australia's most distinguished scientists.

We went by train to Brisbane to join the *Mataram*. Among

the passengers was Raphael Cilento, Director of the Australian Institute of Tropical Medicine in Townsville, Queensland, with whom I discussed mosquitoes. He was later knighted and his daughter, Diane, became a well known actress.

We were met at Gavutu by Major Hewitt, chief of Levers Pacific Plantations, who was largely to organise our survey. We made our working headquarters on the Lever Bros ship *Koonakarra* as we travelled round the plantations. When we spent nights in planter's homes I put on mosquito boots, like soft leather wellingtons, in an attempt to guard against malaria. They, and white clothing, kept us free of disease, before the days of antidotes.

The main problem we had to investigate was premature nutfall of coconuts. There were three ant species involved, one beneficial and two

harmful. We concluded, erroneously, that a bug, *Axiagastus*, was doing the damage. These bugs were plentiful and conspicuous. But the real culprit was shown in a later survey to be a much rarer bug, *Amblypelta*.

When travelling through plantations we were often given horses to ride. At one place I was given the wrong horse, a lively animal which, immediately I had mounted, proceeded to gallop for about half a mile until it stopped in a river to enjoy cooling itself. I am not a good rider and was quite unable to stop the horse in its headlong rush. It must have known it had a complete tiro on its back and took full advantage of this. I had to put up with a bit of teasing when the rest caught up with me.

I had a brief meeting with Garvey at Tulagi, at that time capital of the Solomons. He looked washed out, having suffered bouts of malaria. We had a beer together and he, always an optimist, said no-one should ever miss going to the Solomons.

I collected mosquito larvae and pinned the adults reared from them. Then one day I returned on board, after working ashore, to find my mosquito collection reduced to nothing but the pins. It was a horrid shock. Ants must have been lurking somewhere on the ship and they had stripped my collection. I started again, this time making it ant-proof.

RWP, J.D. Tothill, and C. Widdey (L–R) at Lunga, Solomons, 1928

The highlight of the trip was, certainly for me, and perhaps for Tothill also, the discovery on Faisi, a small island in the extreme west of the group, of *Levuana*-like larvae feeding on a species of wild banana. I was able to rear only one or two adults from the larvae. They were, in fact, *Leptozygaena gracilis* which, years later, I found defoliating a tall pandanus at Bulolo in New Guinea. This zygaenid is the nearest relative to *Levuana* yet to be found outside Fiji, and feeding not on coconut but on *Musa* spp. (bananas and plantains) and *Pandanus* palms.

Back in Suva, Tothill wrote his report on Solomon Islands investigation and asked me to complete my sections of our forthcoming book on the *Levuana* campaign. I was due for home leave, but didn't leave Fiji until the end of November, so my first leave in England after four years was through the winter and early spring. I was thrilled to see trees without leaves and went about Norfolk taking photos of them. It was when the spring foliage was beginning to show that I got the following cable from Suva: 'Taylor finds *Tirathaba* to be an insect of considerable economic importance please proceed to Java earliest opportunity where letter of instructions awaits you Buitenzorg (Bogor) you will study biological control of *rufivena* with view to bringing useful parasites to Fiji'.

The search for coconut spike moth parasites in Java

I went to see Guy Marshall at the Imperial Bureau of Entomology (IBE), now the Commonwealth Institute of Entomology. Besides *Tirathaba* (coconut spike moth) we discussed the *Levuana* publication. Tothill and Taylor had to mail their contributions, but proof-reading could be done only by IBE staff as the three authors would each be located abroad. Tothill had been appointed Director of Agriculture in Uganda; Taylor and his wife were still in Fiji, but shortly also to go to Uganda.

I travelled overland to Genoa where I joined a Dutch ship

Coconut spike badly damaged by Tirathaba

sailing to Java. It stopped to refuel at Sabang, a coaling port at the northern tip of Sumatra. This was my first impression of the Dutch East Indies.

At Buitenzorg [Bogor] I was given a room at the 'Instituut voor Plantenziekten', where Taylor had been working in 1925. The Dutch Director, Dr S. Leefmans, and his staff of entomologists, many of whom spoke good English, greeted me with kindness. I enlisted a native assistant called Djuned, with whom I could converse only in a limited 'pasar' Malay. Djuned possessed one special physical feature, an inch-long little finger nail. Such a 'claw' is grown by Javanese as a mark of social distinction, but for a lab boy who had to clean out glass tubes fouled by the silk and frass of *Tirathaba* larvae, it was an ideal instrument.

I found *Tirathaba* very widely distributed and abundant on coconut flowers in Java. There were two distinguishable

RWP with two Javanese assistants at the Instituut voor Plantenziekten, Bogor, 1931

species, later to be identified as *T. rufivena* and *T. mundella*. The former was the more abundant, but both species were cannibalistic, so had to be kept in separate glass tubes for the emergence of any parasites.

After ten weeks collection and study of the Tirathabas, 17

species of parasites had been reared, a few of which were hyperparasitic. Perhaps after the *Levuana* experience I was prejudiced in favour of Tachinidae, which seemed almost always to be primary.

There was much that I found strange and fascinating in Java. In many houses there

Breeding apparatus for Tirathaba

were rather large gecko lizards known as 'tokeis'. Every so often one heard their 'chuck, chuck ... chuck' in ever rising notes, until at its peak the animal seems to expire its breath in a series of 'tockeis', which after six or so, get fainter until the whole performance is over. These lizards usually hid themselves near the roof timbers and only occasionally could be seen. I made no study of them nor read about their habits, so did not know whether their musical offerings were to do with sex communication. Other smaller gheckos clung to walls in exposed situations and caught moths and other insects attracted to a light on an open verandah. There were also occasional swarms of termites, which flew in at dusk and shed their wings over the floor, which made quite a mess. I stayed at the Hotel Bellevue in a room facing the 7000 ft volcano, Gunung Salak. *Tirathaba* does not fly to a light, but many moth species arrived on the verandah after dark, including the huge Atlas moth, the larvae of which were a minor pest in tea plantations.

My constant companion at the hotel was Dr A. Reyne, one of the entomologists at the Instituut. He told me of a German professor who requested a local scientist to send him six skeletons of the *Pithecanthropus erectus* from Madjun in central Java as if they were lying about in hordes. An American I met said he was going to search for more of this much publicised fossil.

I spent a few weeks visiting central and east Java to check on the distribution of the *Tirathaba* parasite complex. Near Yogyakarta I spent a day walking about the small

Part of the Hindu temple at Borobodur

RWP at the edge of Gunung Gedeh's crater

galleries of the Borobudur Hindu temple. Reyne had persuaded me to purchase a small stereoscopic camera, like the one he had. These cameras were not expensive and one could buy the stereo films for them, but these were mostly glass plates, which are heavy to store in any quantity and a bit clumsy to use. For photographing the stone carvings of the Borobudur temple they were excellent. The temple has a beautiful setting among the paddy fields edged by graceful coconut palms leading up to volcanic cones, dominated by Sumbing (11 000 ft) ten miles distant.

I continued eastwards by train to Banyuwangi, a small town on Java's east coast, separated by a narrow strait from Bali. Dr Loos, local Head of Agriculture, took charge of me and provided transport for the collection of *Tirathaba,* of which there were plenty. Having got enough larvae there were two or three days still available. This was enough to climb the neighbouring volcano, Gunung Raong: 11 700 ft, with a small active vent, continuously roaring (hence its name) in the centre of a vast original crater almost a mile in diameter. I was told of a daring Dutchman who was lowered in a basket down the old crater wall, but when half-way down he was almost suffocated by sulphurous fumes and was hauled up again.

My visit to Gunung Raong involved a bivouac at 9000 ft until 2 am and then a climb to reach the crater edge at dawn, to watch the sun rising gradually illuminating in a glowing, bronzy red the west side of the crater wall. I was spellbound for half an hour and, when the whole crater was sunlit, I took a photograph. By 6 pm I was back in the hotel, although very footsore. I was amused by an advertisement in Dutch and Pasar Malay, 'Goeden morgen. Hebt gij Pears Zeep gebruikt?' and 'Tabeh apa soedah pakah saboen Pears?'

I worked at Banyuwangi with a Javanese agricultural officer called Sumardi. He spoke just enough English for me not to try on him my broken Malay. We did have some minor confusion, because he said 'yes' when we would say 'no' but I learnt to distinguish his *yes* nos from his *no* nos.

One night I carelessly left an apron string touching the wall behind my work bench. Ants demolished 300 *Tirathaba* larvae. To avoid such accidents one needed to be constantly vigilant.

Taylor, still in Fiji, was breeding a species of thrips which Simmonds had brought from Trinidad for control of the pasture weed *Clidemia hirta* (Koster's curse). After that he was to come to Java and take back the parasitic *Apanteles* wasps on an Indian coolieship *Ganges*. This was good news as I wanted to discuss with 'Taggers' the relative merits of the *Tirathaba* parasites, as he had studied the host in Fiji.

Shortly after this I went down with malaria, almost certainly from mosquito bites in East Java. This was the first attack of tertian fever I had. What an odd disease it is. I was in the laboratory one morning and began to shiver. I got in a 'delaman' to drive back to the hotel, piling on all the garments I had available, but was unable to stop shivering.

After about two hours my temperature shot up to 106°F, I became delirious and would gladly have ceased to remain alive. I became bathed in sweat and was given massive doses of quinine, which may have contributed to my later deafness. My temperature soon fell. I felt washed out but otherwise normal. The next day I was back at work.

The Dutch method of dealing with malaria was to give large doses of quinine over four weeks; then a reduced dose over the next two months. This was quite effective and I have never had a recurrence of the disease.

The Taylors arrived on the 26th of December so they had a month in Java before taking my *Apanteles* to Fiji. They brought a small gramophone so I listened to Beethoven and Earnest Lough. Music does wonders to instil a sense of tranquillity.

We all went to the Nichols' tea estate east of Bandung. After a night with these delightful people whom I had met on the *Hooft* we boarded a Fokker aircraft and flew back to Batavia (Jakarta). It was my very first flight in an aeroplane. The plane flew slowly at low altitude so one had an excellent plan view of the west Java countryside.

The Taylors were in Java only six weeks, but it was the first occasion I had really got to know them as a happily married couple. No one else I have ever known made me laugh as much as did Taggers Taylor, with his wit and arch references to the behaviour and actions of the people we had everyday dealings with. Such minor incidents as having tea with Sammy Leefmans, who had a white spot on his nose like that on a Dunhill pipe; finding a perfume in a chemist called 'Po Ho'; and watching a native travelling salesman with his 'pikoel' containers filled with sundry eatables at the end of shoulder-borne bamboo rods, chanting what sounded like 'Ho-Aunty'. Taggers had me convulsed in laughter by his descriptions of such minor happenings.

The Taylors left Java on the SS *Sutlej*. They had inoculations for cholera as there were once again cases on this Indian coolie ship.

Taylor had been studying the coconut leaf-mining beetle *Promecotheca reichii* in the Lau islands of Fiji and left me his report on it. I expected to get asked to seek parasites for it from a related hispid in Java coconuts. Meantime I confined my attention to *Tirathaba* tachinids, the more common of which had been determined by the BM as *Erycia basifulva*. Gravid *Tirathaba* females kept in glass tubes laid eggs in loose balls of cotton-wool and this enabled the easy rearing of the scelionid egg parasite *Telenomus* which seemed well worth taking to Fiji. The only parasite reared from *Tirathaba* pupae was the chalcid *Antrocephalus* sp., itself a potential hyperparasite on the puparia of the tachinid, so ruled out on that account.

It had been rather wet and windy in Bogor and, having decided to get a bicycle to get to work faster, I bought a 'payong', a waxed umbrella, to keep off the rain as I rode. However, a gust of wind blew the payong inside out and landed me and my cycle in a ditch.

Parasites of coconut leaf-mining beetle

Instructions to study parasites of the coconut leaf-mining beetle *Promecotheca* duly came from Fiji and I conducted a search near Purworejo where *Promecotheca* was reported to occur. There was none of it in west Java. Before I left Bogor, Dr R. Aldrich-Blake, botanist from Oxford on a sabbatical, took colour photographs of my index collections of mounted parasites. Prints of these were useful for ready-to-hand reference when the boxes themselves were not available.

The Hotel Bellevue attracted several visitors of interest. Mr Gillette of razor-blade fame was one. Dr Montague, retired chief medical officer, came from Fiji. He told me the sad news that the *Lady Escott* had sunk in a local whirlwind. I had had several pleasant trips on the ship.

To get to the site for collecting the leaf-miner, I flew to Semarang, then took a taxi to Magelang. It got a puncture and had no spare wheel. I found my intended hotel was full and went on to Borobudur where I got a bed of sorts in a small hotel. I was supposed to meet a local agriculture department man at Puworejo, but the only vehicle available had to go first to Yogyakarta where I had to spend another night. Next day I had more luck and was shown coconuts with a little leaf-miner and got enough of it to discover that it was parasitised by several eulophid species.

Back at Bogor I found my room at the Bellevue surrounded by a family of Swedish folk, called De Dardel. The father was to be Swedish consul in Batavia, but awaited the preparation of a suitable house in Weltevreden. He had a wife and two school-age daughters and their governess. They all spoke English well and we soon became good friends. Other visitors were Dr Clunies-Ross and wife, whom I had met in Sydney.

Street scenes in Buitenzorg

Another permanent resident at the hotel was an agricultural soil chemist, Tan Sin Houw. An ethnic Chinese, he was born in Java and educated in Holland. He spoke enough English for me to become friendly with him and I found him a useful contact through whose good offices I acquired a small Fiat four-seater for £100. Petrol cost only 11d a gallon, so it was cheap to run. Its steering wheel was devoid of any covering or paint so I wrapped it in an old puttee. I named the car 'Bertie' and felt quite proud. To get a driving licence I had to get some papers from Fiji, and attach a photograph with thumb prints. I had a driving test by a man who spoke no English. He asked me about bridges, crossroads and speed limits. I think he probably got fed up with my paucity of Malay, but he saw I had ability to drive so gave me a licence. It was, and still is, the only time in seventy years of driving I have had to pass a test.

The best collecting area for *Tirathaba* was along the west coast of Java, about 80 miles from Bogor over roughish roads. I wanted to get 1000 4th and 5th instar larvae, half of them for dissection to see whether *Erycia* conflicted with *Apanteles*. The rest would start a lab breeding stock. On my first trip by car I took with me one of the Instituut's Javan entomologists so he could explain to the locals what was wanted and what I would pay.

Pasaoeran Rest House, where I made headquarters, looked out over the Sunda Strait towards Krakatau Volcano, which looked at the time like a small cone in the distance. Its activities were continuously monitored by a team of radio observers and if there was any sign of activity orders were given for the coastal people to move inland. The Dutch were anxious to avoid a repetition of the 1883 catastrophe when thousands were killed by tidal waves swamping the coast.

Local collectors had produced over 900 caterpillars by the second day of my visit, for which they were paid 1 cent/larva. I felt feverish on the drive back to Bogor, but was glad not to get dysentery from unboiled rest-house water. A letter awaited me from Dr Marshall suggesting I transfer from Fiji to make my base at the IBE in London and then work in Java on the insect pests of Ceylon. I replied enthusiastically but nothing came of it as my contract with Fiji was still unfinished.

The De Dardels had settled into their house in Weltevreden, the residential part of Batavia, and invited me to visit them whenever I liked. I remember on one visit being reproved by Mrs D. for, from habit, helping myself to salt. 'You English, you don't season the vegetables when they are cooking as we do.' I learnt the lesson.

In one identification list received from the IBE there was a scale insect I sent from the Solomons called *Heterococcus painei*; poor thing. A number of other *'painei's'* keep appearing on the lists, which shows how many of the insects I sent to the IBE were new species.

After dissecting about 600 larvae I cabled Fiji to say it would be worth introducing *E. basifulva* as well as the *Telenomus* egg parasite. These were to be bred for a shipment later in the year.

'Bertie' began making ominous noises, so I took him to a nice old Dutch mechanic who thought Java had been a much nicer place in 1877. He decarbonised the engine, ground its valves and made the car behave beautifully.

The reception for the King's birthday at the British Embassy was just a squash and many, like me, had nothing but sweat with which to toast His Majesty. I hoped to meet some of the British whom I knew lived in Batavia, but didn't see any. After a night with the De Dardels I was driven back to Bogor in their 8-cylinder Lincoln car. Ambassadors have to do things in style.

A letter from Taggers told how, after their quarantine for cholera, he went down with dysentery and Rene, his wife, besides nursing him, spent four hours each day tending to the *Apanteles*. She saved the colony and earned an honorarium of £50 from the Coconut Committee.

At the Swedish reception for their King I met a previously unknown Trinity College contemporary called Ted Lambert, with whom I later did several mountain walks. He had to know three European languages for his job in the consular service, and had learnt Siamese in Bangkok and was now learning Dutch. Not all Englishmen, like me, are poor linguists.

'Bertie', loaded up with Tirathaba *larvae, on the road back from the west coast to Bogor in 1930*

One evening I got my friend Tan to take me to one of the top Chinese restaurants in Batavia. I wanted to try birds-nest soup, trepang and a special 'bami'. On the way I somehow got Bertie stuck between the gates of a level crossing. We were in no danger from trains, but I had to produce my driving licence and, thanks to Tan's diplomacy, I got away with just a caution. The thing I enjoyed most on the menu was called 'panang babek', roasted duck skin. To counteract the fattiness of it we drank pale Chinese tea.

During a holiday weekend I took the Sydney entomologist, Windred, to the Gedeh and Pangerango summits. At 6 am the temperature there was 39°F. I made porridge, which Windred thought disgusting, but which seemed to me to hit the spot. We walked across the summits and came down on the Sukabumi side. Half way down, hurrying, I realised I had left my field glasses on a tree not far from the summit. There was no time to go back for them that day, but I did recover them again three days later. I set out from Bogor at 4 am and climbed 6000 ft, got the field glasses and was back at Bogor for lunch. It was an unsponsored exploit and good fun.

Introducing *Megarhinus,* a predatory mosquito

I suggested to the Fijian authorities that I could introduce the large predaceous mosquito *Megarhinus* (now *Toxorhynchites*) which lays eggs in water-holding tree holes in the Bogor Botanic Gardens. Its larvae feed on those of other mosquitoes. It should have been possible to establish this mosquito in Fiji, where holes in Ivi trees contain larvae of *Aedes*, vector of filaria and a day-biting irritant in coconut plantations.

Mid-year it was a frequent sight to see a dense grey cloud approaching from Mount Salak accompanied by flashes of lightning and thunder. As it got nearer, rain began to drench the whole of Bogor, with overhead flashes and alarmingly instantaneous deafening thunderclaps. The deluge could last two hours; but often by 5 pm it had stopped, the sun had come out and the ground just steamed.

I was laid up with a feverish chill when *Erycia* flies were due to emerge. I told Djuned to transfer them to a large cage, but some object beneath the cage formed a bridge for ants, which ate all the flies. This was a major setback which meant another collecting trip to the west coast. I had observed, however, before being bedridden, that *Erycia* when ready to oviposit was attracted by the silky, frass-lined tunnels along which the *Tirathaba* larvae feed. The fly pokes eggs along these tunnels with its long flexible ovipositor. The host larvae must either swallow the fly eggs, or the eggs hatch and the fly larvae crawl onto their host. Assuming that the eggs might be swallowed I devised a method of getting *Tirathaba* larvae infected. By placing fly eggs on the moist, freshly cut base of a single male coconut flower and then putting this in a tube with a single caterpillar, the latter would swallow the egg. When this method had been perfected it formed the basis for the breeding of *Erycia* in captivity. It is referred to later.

The English padre in Batavia told me about a Dutch couple whose baby he had christened. Their name was Meerkamp van Emden; they had a house near Bogor and spoke good English. He worked for Shell Oil. She was part Filipino and they were married in the Philippines. They seemed to want their son to go to Eton and wondered if I could help. They also seemed worried that I hadn't got a wife, and wondered how I coped. They were unsuccessful in their attempts to help me in that regard.

I saw advertisements in some entomological journals for our book on *Levuana*, but I hadn't yet seen the book itself. Soon, in January 1931, I received from Fiji a rather battered parcel containing my allocation of six copies. They were rather heavy books and had received scant wrapping, so arrived rather bruised. I was very disappointed. I complained about it but never managed to get any more.

Leefmans invited me to a meeting of the local Entomological Society to talk about my work. I thought a demonstration of the predatory habits of *Megarhinus* mosquitoes might be of interest. I had a jar full of *Culex* larvae into which I put a larva of *Megarhinus*. The latter didn't appear to be hungry and made no attempt to feed. So I put in two more *Megarhinus* larvae which, instead of grabbing *Culex*, attacked each other. So the demonstration was a flop. However, it did, quite unintentionally, show that in the sort of confined breeding habitat of *Megarhinus* it is probably essential that only a single individual matures if they are to avoid being stunted, so competitors are eliminated.

I arranged with Ted Lambert to walk up Java's most westerly volcano, Mt Karang. It is only 6000 ft high, forest clad to the summit, where there was a rather obsolete meteorological installation. On the way I got some flowers of an interesting palm, *Plectocomia*, a climbing rattan. The flowers hang in long streamers with a series of cup-shaped brown bracts. They were used as decoration on a ceremonial arch for Queen Wilhelmina's 50th birthday on 30th August. They are also host for a *Tirathaba (nuptilinea)*.

I must refer again to the method devised for breeding *Erycia* in captivity. I may not have had quite such a thrill as Professor Steptoe and Mr Edwards had with their first test-tube baby, but my test-tube fly babes were to me a much-needed stimulus. I should have published an account of the technique, but didn't and it wasn't until 1956 that B.A. O'Connor, with my permission, published an account of it in a New Zealand journal.

Gunung Salak volcano, seen from the Bellevue Hotel, Bogor

I don't know when international telephone calls were first possible, but I think the Dutch in Java were pioneers of this facility. I was delighted when I was told I could speak to my father on his 60th birthday, December 11th, 1930. I booked a call at 11 pm Bogor time, which was 4 pm in England. I took the call in a phone box and father's voice was very clear and we chatted for four minutes. I felt quite bewildered when I emerged from the box, feeling I had been at home. I was told it was the first call that had been made to England from Buitenzorg.

Sir Josiah Crosby, the British Consul in Java, an oldish bachelor, had just returned from leave. I asked Ted Lambert if he could bring Crosby to dine with me in my new rooms at the Bellevue. I had been allotted a special suite with an uninterrupted view over the river to Salak — the view painted by my Aunt Frances de la Poer in 1898. Now, 91 years later, Aunt Frances's watercolour is in my study.

After a good dinner we sat and smoked on the verandah and talked of fundamentals. Crosby said he didn't believe in free will. All was preordered: we were just puppets fulfilling our destined purposes. It seemed in that case that no-one was responsible for their behaviour. Neither Lambert nor I could agree to that. But, however odd Crosby's views may have been, he offered me help with any shipment arrangements when I took parasites to Fiji. This proved very useful at Noumea, where I gave his letter to the consul.

On Christmas morning I drove to the English church for the early service, had breakfast with the Cribbs and stayed for

Erycia *breeding cages*

At the New Year holiday the Meerkamps drove me to Bandung where I had my first hot bath for 11 months. At Bogor one could only slosh cold water over oneself with a dipper. There was no hot water. I was back on a bicycle having sold my little Fiat. There was no point in licensing it for another year.

matins to sing the proper hymns. The De Dardel family all came, a nice surprise. Returning to Bogor and a hurried lunch and half hour at the laboratory, I drove to Ardjsari to spend the rest of Christmas with the Nichols. I had a batch of letters from Fiji — tales of woe, as they had had two hurricanes, the second one wiping out the banana crop, a serious loss. Taylor wrote that the Coconut Committee were mad at the thought that I had all this time been enjoying myself in Java without much to do and they intended to get me back to Fiji at the earliest opportunity. So it seemed that I had been sitting and gazing at Mt Salak, doing nothing. Ah me!

Shipment of parasites to Fiji, 1931

On 13th January 1931, I heard from Fiji that I could sail by a Dutch steamer to Noumea and there tranship to a diverted steamer, on to Suva. It looked like a long journey during which I should have to breed the insects if they were to reach Fiji in sufficient numbers. Heavy rains had just washed out all the tree holes and there were only two weeks left to procure an adequate lot of *Megarhinus*. During the last week before sailing I procured 52 *Megarhinus* eggs and I had enough *Tirathaba* parasites and hosts to start the voyage with some chance of success. Crosby gave me a diplomatic letter to the French authorities at Noumea requesting facilities for me.

It was an anxious time getting cages on the ship, and seeing none were left for long in the sun. I took Usman, my recent lab assistant, as far as Semarang, where he could catch a train back to Bogor. SS *Le Maire* was a small freight ship with limited passenger accommodation. I had a small, very hot cabin. The insect cages were shaded on deck. At Semarang, in central Java, I collected *Culex* larvae for the hatching *Megarhinus*, carefully kept from each other in different tubes; also coconut flowers for the *Tirathaba* larvae. I could see the volcano Sumbing and thought of the rainy moments I had spent at its summit. Merapi, another volcano, had recently erupted, killing 800 Javanese. Java is a truly remarkable island, fertility the price of risk.

On 6th February we had been a week at sea, with three more days to reach Port Moresby. The heat and vibration of the ship through all this calm sea were preventing matings of both *Erycia* and its host. After Samarai, where fresh coconut flowers were obtained, the weather turned cooler and more windy, but still no matings were taking place and the situation was critical.

Megarhinus alone seemed all right.

The rest depended on getting supplies of *Tirathaba* at Noumea and I had no knowledge of the New Caledonia fauna.

However, at Noumea I found *T. rufivena* plentiful, was able to get its eggs for *Telenomus* and met Jean Risbec, the local entomologist, who arranged for the cages to be put in the shade in the Governor's garden.

The ship *Wairuna* from Sydney, diverted to take me on to Fiji, arrived later than expected and I had to spend five days at Noumea. None of the flies mated at Noumea, but I didn't discover that until I dissected them. We reached Suva on February 27th, just a month after leaving Java. The place was in a dither; the barometer was falling and the wind rose. A hurricane struck the northwest of Viti Levu, but luckily Suva escaped.

Erycia cage on the SS Le Maire

The Taylors met me, also R.A. Lever, a young entomologist, destined for the Solomons, but staying in Fiji to get a few wrinkles from Taggers and me. The Taylors were able to take leave the next month. They had *Apanteles tirathabae* established. The insects I had brought were still alive and some newly emerged *Erycia* had mated. I had also brought an ichneumonid, at that time referred to as *Nemeritis palmaris*, which attacked *Tirathaba* larvae at a later stage than *Apanteles*, but did not conflict with *Erycia*. If they all got established there should be a fair degree of control of the spike moth. *Megarhinus* survived the journey well. I found lodgings in a house belonging to the widow of a jeweller called Levy. It was quite a long way

Unloading the cage in Fiji

from my insect cage, but I bought, for £130, a Ford two-seater which had only done 9000 miles. I found Mr Barnes, the new Director, helpful and easy to get on with.

One evening, when supping with the Taylors, Hop Ackland telephoned to say a hurricane was on its way and if it hit Suva the parasite cage would be blown out to sea. We had to move it indoors. All that night and next morning the wind blew strongly. At 11 am, Barnes ordered the Department to be evacuated and the shutters put up. It rained very hard but the hurricane passed 80 miles east of Suva. During all this scare one of the mated flies died before its eggs were ready for use. Two others mated the following day, but it was touch and go whether the breeding stock would survive. My life at this time was on tenterhooks.

After the hurricane scare was over the public came to 'see the bugs'. The large mosquito was a particular attraction and when they had seen it feasting on noxious mosquito larvae tales got round that it would eliminate the mosquito plague. I had to impress on people that *Megarhinus* laid its eggs only in tree holes, so could have no effect on the bulk of mosquitoes which breed in other water containers. I hoped that it might reduce the incidence of filaria by feeding on such of the vectors which bred in holes in trees. The new Governor, Sir Merchison Fletcher, also inspected the insects and a reporter from *The Fiji Times & Herald* asked for my impressions of Java, which I hoped would not be too mangled when printed.

I hitched a lift to Bengga with a

party of Americans eager to see the Fire Walkers. I released *Nemeritis* there. The first colony of *Telenomus* was released in the Rewa delta where it was nice to see coconuts bearing good crops again after the control of *Levuana*. I went on to Bau to call on Ratu Pope. Sukana and John Goepel were there, the latter a Cambridge graduate, competent piano player, and well versed in G & S operas. Pope gave us the tastiest Fiji meal I'd ever had — oysters, snipe, fish, and that delicious 'vokalolo' pudding made with cane sugar and the milk from coconut.

The *Erycia* situation remained critical. I prayed in church for their survival. It seemed to have worked as I got two mated flies and was able to prepare another 500 host larvae for infection. I was anxious to get these to Taveuni as quickly as possible. Barnes said he couldn't afford to fly me there, so I went by launch, taking about 300

infected *Tirathaba* larvae and hoped that the emerging flies would be enough to get the species established. The release was made on dwarf coconuts at Mua. On one of the leaves there was a nest of hornets which began to attack me. I grabbed a green frond and descended with its aid to the ground, a stunt I would never have done in cold blood.

Back in Suva I met Ian Ramsay, proceeding on leave from Java and anxious to see the south seas. I had got to know him in Java, where he worked for MacLaine Watsons. His father, an Islay laird, was tutored for Oxford by my father, so I knew something of the family. Iain was a skilled yachtsman off the Scottish coast and I had him to stay and took him on a Fiji cutter, chartered to take parasites to Suva and Koro. Ramsay, who wore a kilt in the evening, performed a sword dance at Government House, but to do the dance justice he had to

perform barefoot. This rather shocked the Governor, or perhaps his aide, and I got a reprimand for allowing such crude behaviour.

By early June the weather had turned cold and the last of the fly breeding stock had died unmated. So establishment of *Erycia* depended on those released at Mua.

The financial crash of 1929–30 had its effect in Fiji. I could no longer pay a Fijian £3 a month for breeding *Megarhinus*. As several had been released near Suva and also in an ideal situation on Taveuni, it became established.

I spent a month on Vanua Levu and Taveuni. On Taveuni I stayed in the overseer's house at Mua, where, for the first time, I was attacked by bed bugs. As a precaution I had my baggage fumigated on returning to Suva. I could find no *Apanteles* on Taveuni, although it was well established around Suva.

In September I took parasites to the north coast of Viti Levu. It was a relief to escape from all the social activities in Suva, where two New Zealand naval vessels were in port. After releasing *Telenomus* and *Apanteles* I walked up to Nandarivatu to spend a night in the cool. The next day I started to walk overland to

Fijian village bures

Vunindawa. As this was 60 miles away, a bit much for one day, I spent a night in a pretty village, Nasogo, where I supped off yams and dalo, staple food for Fijians, but which I found too starchy, although very filling. I reached Vunindawa by moonlight the following day and called on District Officers Jack Bye and Jimmy Windrum, recent friends. They plied me with whisky and fed me with four fried eggs, followed by a hot bath and a comfortable bed. Great hospitality. The next day I got a launch down the Rewa to Nausori and a taxi back to Suva.

Telenomus had been recovered at Lami across the bay from Suva, and it appeared that the little parasite should get established all right. In mid October I moved camp to Taveuni. I took our *Levuana* assistant, Jacob, with all the gear for breeding the three hymenopterous *Tirathaba* parasites. I got digs in the rest house at the

Government Station, Waiyevo, which had an Indian cook.

Bob Snodgrass, a stumpy Scotsman, was both Medical and District Officer there. His wife, Viti, was the eldest daughter of Mrs Mackenzie, known as 'Queen of the North', the owner of Nagasau coconut plantation.

Beveridge, an assistant from the entomology division in Suva, and his wife, were to take over the parasite breeding routine and would have to occupy the Rest House, so Bob and Viti asked me to stay with them.

Shortly after my arrival on Taveuni the *Malaki*, a Burns Philp steamer, called on her way to Rotuma, an island dependency of Fiji, some distance north of the group. It seemed a chance to release any available parasites there, as well as *Megarhinus*. Rotuma is a very fertile island renowned for the large coconuts borne on the palms there. A fellow passenger was Mr Craig, the Colonial Treasurer. We played piquet together, a game I had only previously played at home. It's an excellent card game for two, but required more luck than skill. There was little time to see much of Rotuma. The Resident Officer was Dr Carew, a pleasant and imposing person, who made small golf courses wherever he was stationed. His son Desmond was a good friend of Nott's.

A hyperparasite had appeared attacking our *Apanteles*. This was a pity, but made *Erycia* all the more important. It was necessary to find out whether the 300 infected hosts released at Mua early in May had established the species there. There had been time for several generations, so it should have been recoverable if it had continued to breed. With three Fijians climbing, some hundreds of hosts were obtained from which to rear the parasite. Sadly, not one

produced *Erycia,* which evidently had failed to establish. Obviously it wasn't going to be as easy to establish as *B. remota*, the *Levuana* tachinid. Conditions in Suva after the hurricane scare made the breeding of the parasite much too hazardous. It was too cool. If the Coconut Committee had wanted to get this parasite into Fiji it would have been best to set up a breeding station on Taveuni where there was shelter from the cool southeast trade winds.

I was invited to spend Christmas with the Mackenzies at Nagasau. There was a large house party including the MacKays from the adjacent plantation, Nathaugai. I shared a bedroom with young Max who began his day by tucking into a box of chocolates. I refused them, disliking chocs at that hour. I survived the remaining Xmas festivities. Returning for the New Year I narrowly missed a four gallon douche of water which had

been fixed to descend on me as I arrived in the morning. Luckily for me, one of the plantation staff arrived first and got the showerbath. A thing none of us escaped was the 25 inches of rain which descended on the island during January 1932.

Barnes told me I could take long leave in March 1932. He thought neither the spike moth nor leaf-miner projects would be finished before 1934 when my contract was due to end.

During February Beveridge and I did several parasite releases in islands of the Lau group. At Munia, in the early morning I hooked a large tuna while rod fishing. It took half an hour to play and haul into the boat and weighed 70 lb.

Home leave, 1932

I was hoping to proceed to England through the Panama Canal, but the value of sterling was depressed and I was told I would have to take a steamer from Dunedin and sail round Cape Horn. Looking on the bright side I was glad to get a chance to visit New Zealand. I spent a day at Rotorua and enjoyed a long wallow in the thermal pool at Wairakei. After Java's volcanoes in the tropics it was interesting to compare the snowclad Mt Egmont and the volcanic pair, Ruapehu and Ngaurahoe.

One couldn't see much of the South Island mountains from the train en route to Dunedin, but I found the Otago University Museum well worth a visit, as was a lecture by Dr Skinner at a meeting of the Archaeological Branch of the Otago Institute.

The SS *Port Napier* took 21 days to reach Montevideo. We seemed to bowl along day after day on the fringe of the roaring forties, with little to watch except the effortless gliding of albatrosses as they followed the ship — what glorious birds they are. There were only eight other passengers, among whom Dudley Atkins, a solidly built, middle-aged Englishman, resident in New Zealand, became my companion. Our skipper was rather new to the job and gave Cape Horn a wide berth; but the weather was so misty and gloomy we would have seen little of it had the ship been closer.

At Montevideo, Atkins had a cousin with a daughter called Inez. He invited me to spend the day with them. He thought that I, being a young bachelor, and Inez a young spinster, might fall in love. I'm afraid he was disappointed. My thoughts were of a British lass. Northwards through the tropics there were deck games, in which two wool buyers, one German and one French, were keen players and displayed diverse responses to being beaten.

The ship called at Las Palmas to bunker and eventually docked in London on 5th June. How nice to have the prospect of a summer in England, or more likely Scotland, but the voyage had lopped 45 days off my leave.

I went to have discussions with Guy Marshall at the IBE. He mentioned possible jobs in Mauritius and Solomon Islands. I told him I was anxious to get *Erycia* established in Fiji, although the extension of my contract had yet to be confirmed. Next I went to see Dr W.R. Thompson at Chalfont St Peter. W.H. Thorp, a fellow graduate at Cambridge under Balfour-Brown, was working for Thompson in the Imperial Institute of Biological Control. I felt a bit frightened of Thompson. He was a formidable mathematician, whose publications were always laced with algebraic formulae, to me both obscure and somewhat daunting. I held the view that biological control involved studying living insects in their natural environment, not producing models to predict their hypothetical behaviour. I always regarded Taggers Taylor as a good example of an effective entomologist, in all of whose field work mathematics was only once used. That was when he had, in a matter of weeks, to make an assessment of damage done by *Tirathaba*, an assessment which subsequently attracted some criticism.

Later on, Thompson moved his institute to Ottawa. There his insect population dynamics were challenged by A.J. Nicholson, whom I had met in Sydney. The antagonism between these two entomological pundits grew with time. Who won in the end I never knew as my career as an entomologist was interrupted between 1935 and well after the war.

Later, during the summer of 1932, I called on Mr Stockdale in Whitehall. He was liaison officer between Guy Marshall and the Colonial Office. He told me that Fiji was unwilling to release me and there was no need to seek another job.

In October I met Evelyn Cheesman at the BM. It was my first introduction to this splendid South Pacific insect collector. She was still working through her collections when I met her again in 1956. On that occasion she gave me a rough map of the Banks Is. which I hoped sometime to visit. I said 'Won't you still need it?'. She got off her stool with a stick, gave her thigh a slap and said 'That won't let me do any more field work.' Her biography 'Things Worth While' (Hutchinson, London, 1957) is worth reading.

I started on a paper, 'The Introduction of *Megarhinus* to Fiji'. I called on F.W. Edwards at his home in Letchworth. He had been my BM contact for mosquito determinations. He was a friendly and helpful person who collaborated for a paper on mosquitoes from the Solomons.

Early in November I received word from the Colonial Office that I was to return to Java to get another lot of *Erycia* to take to Fiji. On another visit to the BM, I overstayed the staff, got locked in and had to escape through a window. The time for such free and easy going is, alas, no longer — at least from a visitor's point of view. Security is now very strict.

During the last few days in England I saw Pat Buxton at the London School of Hygiene and Tropical Medicine, got up to date with anti-malarial measures and completed the *Megarhinus* paper. I invited, for lunch at the Russian restaurant, Marshall, Sam Neave and Arthur Cobb. Arthur thought I shouldn't be contributing to the increase of tropical food crops as he considered there was a surplus of them. So he and Marshall had an argument about that, while Neave and I listened, amused by Arthur's rather parochial ideas.

Return to Fiji, via Singapore and Java

I joined the P & O *Carthage* at Marseille. It was a memorable voyage for me as I got engaged to Noel (Fawcett). She was on her way to stay with an uncle, colonel of a Ghurkha regiment at Peshawar. I bought her a ring in Bombay where we had to part company. The uncle was a bit taken aback when his niece arrived with an engagement ring on her finger and warned her about the instability of shipboard romances. Noel and I were to show that the reverse is sometimes true: we went on to celebrate our golden wedding anniversary.

I had to spend a week in Singapore, as the mail steamer to Java had broken down. I visited the strikingly white Anglican Cathedral and asked the Archdeacon if Noel and I could get married there in February. He raised no objection, as we each had British passports.

I sailed to Batavia on the small steamer *Ophir* which arrived on New Year's Day, 1933. The Taylors were staying at the Hotel Keyzer on Jalan Pledang in Bogor, so I booked in there. It was nearer the Instituut than the Hotel Bellevue, which had been taken over by government officials. We had to speak Malay to our hotel management. Taggers was working on coconut leaf-miner parasites, breeding up a stock of two eulophid parasites which he hoped soon to take to Fiji. I hired a car and went to the west coast to get as many large *Tirathaba* larvae as possible. The old man who kept the Pasaoeran Rest House remembered me and gave me back a hazel walking stick I must have left there.

Anopheles mosquitoes were abundant. I wore mosquito boots, buttoned-up my white shirt sleeves and took quinine tablets. Local collectors soon had the 300 larvae required. There was some excitement when Krakatau started to

pour out a column of smoke which rose vertically in the still air. After dark, fireflies were a lovely sight, gently flying to and fro with a thin pale-green light pulsating as they flew.

After returning to Bogor I had a mail from Peshawar. Noel said she could get a passage to Singapore in February. I cabled Fiji asking for two weeks leave to get married in Singapore. I said that I would arrange for the breeding of my insects to be supervised in my absence. Mr Barnes cabled approval, for which I was duly grateful.

Arrangements for our wedding were greatly helped by Noel's father discovering that Charles Wurtzberg, a friend, would accommodate and give away his daughter. For my part I enlisted as my best man Percy McIlwaine, formerly Attorney General, Fiji. We were married in Singapore Cathedral on 21 February 1933, which also happened to be my thirtieth birthday. We flew back to Java early the following morning. In an old Ford car we spent our short honeymoon at Bandung and Garut in the pleasantly cool elevated conditions where we could enjoy a hot bath and listen to records on our new portable gramophone.

Back at Bogor, we could also enjoy having Tagger's and Rene's fairly mature marital state to reaffirm the delights of our plunge into such an existence. It was a hilarious few weeks before we were to depart together for Fiji.

Noel and I paid a last visit to the west coast for additional *Tirathaba* larvae. These would yield flies which could be used to infect host larvae during the early stage of the voyage. With the Taylors and their leaf-miner parasites we all

The 'wedding breakfast' in Singapore, 21 February 1933

left Java on 17th April by the small KPM steamer *Van Rees*. We took as far as Semarang our lab assistants, Usman and Udjyang. It was extremely hot in Semarang where the hatches had to be opened to take on a load of sugar. Some newly emerged flies died from the heat, but this was the sort of hazard I had grown used to on such journeys. We had reserves, so could shrug off the loss without undue anxiety. At Surabaya there was the last chance to cut fresh coconut flowers until reaching Port Moresby in a week's time.

For the following week the ship sailed through a glassy sea, with the mountains of Bali, Lombok and Flores on the starboard horizon. Our charges seemed to be happy and the Captain, Chief Officer and Engineer were both obliging and entertaining.

At Port Moresby I fed a new batch of *Tirathaba* larvae with eggs of *Erycia*. This delicate operation beneath a binocular microscope would have been impossible at sea. At Rabaul, the agricultural chief, Mr Murray, drove us to Keravat, a recently started experiment station, where Froggatt and Green were the entomologists. Murray said that a local species of *Megarhinus* bred in coconut husks. It had been

Fly parasite breeding cages at Bogor

taken by C.E. Pemberton to Hawaii but failed to establish itself. We searched for this *M. inornatus* species, smaller than *M. splendens*, but couldn't find any. However, a local doctor was breeding it and gave us some to take to Fiji. So far as I know, it has never been recovered there.

On May 8th we arrived at Vila, where we were able to get a supply of fresh coconut flowers. On the next day we reached Noumea. Here we had to transfer to another ship as the *Van Rees* was going on to Sydney. On the *Karetu* the insects had to be kept on deck with a tarpaulin to keep off

salt spray. It was a small ship, very uncomfortable, but it took only two days to Lautoka, its first port of call in Fiji. The Taylors, with their small insect cages, were able to proceed by launch to Suva. We had to wait till the *Karetu* was ready to sail to Suva with our large fly cage (6' x 4' x 4').

At Suva the office was very congested and we had to wait 11 days before sailing to Taveuni on a launch only just large enough to take us and the insects. It was a rough overnight trip to Vuna, not the most pleasant introduction to Fiji sailing for Noel. But there we were taken to an unoccupied bungalow on the Ura Estate, put at our disposal. Our new home was called 'Wawatheva'. We had a very happy and productive 14 months there.

I signed on two Fijian assistants in Suva and bought a two-seater Ford tourer. The assistants, Ilaitea and Sivarosi, had a small shack put

Noel at 'Wawatheva'

up for them, and the fly cage was kept beneath an open coconut leaf shelter, protected from rain, but with dappled sunshine. So far as fly survival was concerned, only one usable female was obtained from 37 adults which survived the voyage from Java, but that was enough to get a breeding stock started on Taveuni leading to the establishment of this most efficient parasite throughout Fiji. It does demonstrate the sort of knife-edge situation applicable to some parasite transport and initial breeding.

Tirathaba Campaign staff at Waiyevo, Taveuni; RWP at the far left.

The stock of *M. inornatus* was released in tubs at Ura. *M. splendens* was well established from the 1931 introduction, not only on Taveuni, but in parts of the Lau group where I had taken it.

From a neighbour we bought a Sydney Silky terrier bitch. Early one day she was barking furiously at a large longicorn beetle *(Xixuthros heros)*. It was nine inches long, including antennae. It was nice to get a specimen of one of the world's largest beetles.

In December, A.C. Smith, a young American botanist, visited Taveuni. We had him to stay. I escorted him into the forest and to the summit of Ulu-i-galau (4030 ft), Taveuni's highest mountain. I showed him the famous 'tagimaucea' flower, always said to be an orchid. Smith told me its true identity, a melastomatid plant (*Medenilla* sp.), a relative of *Clidemia hirta* (Koster's curse) one of Fiji's worst weeds. However, this somewhat inferior systematic status for such an attractive flower did nothing to detract from its value as an advertising symbol for the tourist industry that later developed in Fiji, as I discovered in 1976.

We were invited by Ken Allardice to spend Christmas at his estate in Lau, where the Taylors had made headquarters for their breeding and distribution of the leaf-miner parasites. Nambavatu estate was mostly on a raised coral platform at the edge of which one could look out over a bay studded with coral islets. Ken, whose brother had married a friend of my family, was a well educated Scot. He had lived a somewhat refined existence for long enough at Nambavatu to acquire fluent Fijian. Taylor was getting excellent results from one of the Java parasites, which eventually gave complete control of the Fijian leaf-miner.

I walked through the island to Loma Loma, the government station, where I spent a night with Sukuna who was District Commissioner there. Noel and I later called on the Hennings family at Naitamba, where it was always a joy to see their beautifully kept plantation.

After the New Year the Taylors spent two nights with us at Ura and took a colony of *Erycia* to release at the north of the island. Al Smith came to stay again and during a walk in the bush with him I captured a female *Cotylosoma*, a phasmid, which at that time was considered aquatic. It had leaf-like lateral projections from its thorax thought to be gills. My specimen was taken on a tree trunk. It laid a few eggs before it died. Later I captured a male and sent the material to the BM where Boris Uvarov wrote about them in a paper entitled 'The Myth of Semi-Aquatic Phasmids'. Years later Mahdu Kamath, an entomologist friend, captured

several *Cotylosoma* of a similar species on Viti Levu, showing that it is not confined to Taveuni as was previously thought.

At Wawatheva we were becoming increasingly nauseated by a most unpleasant smell. So disgusting was it we had to move from our bedroom to a remote part of the verandah. Where the smell was most powerful we noticed a drip from the ceiling. The removal of a board revealed a putrifying, five foot long snake. The remains of this decaying corpse were dumped in the sea and the house disinfected.

After releasing colonies of our tachinid parasite at several places on Taveuni and some nearby islands, I boarded a native cutter going to Kandavu, an island in the south of Fiji. Besides *Erycia* I released *Megarhinus* there. I took Sivorosi, whose home was on a small island off Kandavu. The 'Buli' (officially appointed head man) of the village was an albino. Albino Fijians always look odd with their pale freckled skin, but there seemed no stigma attached to them.

Over an area not far from our house there were large blocks of an ancient lava flow between which there were narrow paths. Towards the end of March we were walking along one of these, hoping to shoot pigeons. Pickles, our dog, ran ahead of us and disturbed a large nest of hornets. In such a restricted situation it was impossible to avoid them. Noel got fifteen stings. I got only three. After retreating from the nest we removed top garments to shake off any remaining hornets. Our bare skin became nearly black with mosquitoes, but we couldn't feel their bites with all the pain from the hornet stings. Apart from the intense irritation, an after-effect of the stings, Noel survived better than I did. She must have been less susceptible to wasp venom than I was — certainly braver.

We had Wal and Dorothy Warden to dine one evening. They had become our closest friends on the island. It all stemmed from Wal's experience of convalescing after a wound in the 1914–18 war. On that occasion he stayed with an Etonian family called Harris. They had a large

Gus Hemmings and an Erycia *release cage, 1933*

house in Cumberland. Wal received lavish attention from the Harris family and by befriending Noel and me he felt he was repaying a debt. Shortly afterwards, Noel went to stay with the Wardens at Songgulu to be within easier distance of the Cottage Hospital at Waiyevo, where she expected our first child to be born. It was as well she did. There were intervals of heavy rain during May, and although Noel was able to get to the hospital in time for the birth, I was prevented by flooded creeks until the second day after our baby son, Christopher, had been born. The rainfall at Ura for the twelve days to the 28th of May was over 25 inches. At Salialevu, the wettest estate on the island, 70 inches fell. Christopher seemed a well chosen name.

After several more *Tirathaba* parasite releases on a number of islands I handed over supervision of the breeding work to Beveridge and, on

Fellow travellers on the Niagara, *1934*

17th September, we returned to Suva. We were invited to stay at Dr Jack's house. He was the newly appointed Director of Agriculture, whose wife hadn't yet come to the colony. During the two months we spent in Suva I completed

a paper on the spike moth as well as a short one on mosquitoes.

We sailed for England on the RMS *Niagara* on November 16. A fellow passenger was Arthur Bruin with whom we formed a life-long friendship. We were met by C.E. Pemberton in Honolulu. He had become the chief of the Hawaiian Sugar Planters Association (HSPA). At Vancouver, my relations at Alberni, V.I. took charge of us, had us to spend a night, and put us on the Canadian Pacific Railroad train eastwards. Harry Ladd had invited us to visit them at Rochester, so we entered U.S. at North Portal, where for a time we were in a different part of the train from Christopher. It was a momentary hitch, as was our attempt in Chicago to get him looked after at a hotel while we explored the city. Nothing doing over that. After some enjoyment with the Ladds we proceeded to New York and boarded the *Berengaria* for the last lap of the journey. Mabel Wayne, a fellow passenger and composer of lyrics, took a fancy to Christopher and composed a hit number 'His Majesty the Baby'.

I arrived in England, after almost ten years of tropical life, convinced that biological control was the most satisfactory way of dealing with insect pests, and the only type of entomology that really appealed to me. At the same time we also wanted our future base to be in U.K. There was no vacancy in the 'pool' of entomologists attached to W.R. Thompson's Commonwealth Institute for Biological Control (CIBC), and the only form of biological control used in Britain at that time was for two glasshouse pests. In 1934 there was no organisation devoted to the control of field pest insects by biological methods. I did go for an interview at Cambridge for some proposed job, but that demanded a knowledge of chemical control agents, so I wasn't considered for it, and didn't really want it.

INTERLUDE — 1935–1955

A market gardener in Scotland

After meeting Noel's parents, we spent Christmas with my family at their home in Norfolk; then we took up an offer from my sister and her husband to occupy a house they were having built at Onich, a small village in the West Highlands of Scotland, a place at which my family had spent summer holidays. While there I was more than a year exploring various possibilities for work, mostly with some horticultural basis, as it seemed that, if entomology had to be abandoned, a future connected with growing plants had strongest appeal to both of us.

During 1935 a second son was born to us. We called him Sandy.

Friends and relations living in the Fort William area complained about the lack of fresh vegetables in local shops and suggested that we might try to supply that want by starting a market garden in the area. We both felt strongly attracted by this proposition. What we lacked was sufficient capital and the specialised skill to get the scheme under way. But with the idea firmly rooted in our minds we began to explore the possibilities.

We got in touch with the regional horticultural advisory services, searched the whole Lochaber area for a suitable site for a market garden and arranged with a relation for a loan of one thousand pounds to get the project started.

Although Onich comes under the North of Scotland College of Agriculture in Aberdeen, climatically it seemed more sensible to seek advice from the West of Scotland College in Glasgow. An adviser from that College came to see us, as well as to inspect the few sites which we thought might be possible for market gardening. He not only agreed about the site we thought most favourable, but put me in touch with two very able practitioners of commercial horticulture: Robert Scarlett of Musselburgh and James Warnock of Garrion Farm in the Clyde Valley. Warnock offered to have me as an unpaid apprentice at Garrion, so I got accommodation for us at a farmhouse near Strathaven in Lanarkshire.

It had been impressed on me by the College adviser who had seen the site at Onich that it would be essential in that climate to have glasshouses in which to grow tomatoes as a main crop. At Garrion Farm there was a range of glasshouses, so it was possible to see the various stages of market-garden, nursery and field cropping including the crops we intended to grow at our Highland site.

View from the future house site at Corran, with Loch Linnhe and the steamer to Oban in the background

By the end of 1936 we had obtained the 'feu' of our land. A feu is a lease in perpetuity, so long as an annual duty is paid. It gives full control of the use of the land, but without mineral rights, and subject to the approval by 'the Superior' of any buildings erected on the site. The site we selected consisted of four acres of arable land, cropped in traditional crofter rotation, and 20 acres of rough grazing. The whole area was bounded by Forestry Commission plantings. Geologically, it was a portion of an old glacial outwash terrace. It had only 4 or 5 inches of top soil of a very acid nature, overlying a packed gravelly subsoil. One could hardly call it ideal land for productive market gardening. Nevertheless, it was the best available and, at least at Corran, the name by which our site was known, a market for produce was less than ten miles distant in Fort William.

Just as we had made all arrangements to start on our

venture a cable arrived from Dr Jack offering me the job of Government Entomologist in Fiji, on the early retirement of H.W. Simmonds. This, I must admit, was rather a teaser. I had to refuse. I didn't want to go back on our decision to avoid further living in the tropics. Thus it happened that I changed from being an entomologist to becoming a market gardener. But it wasn't to last for the rest of my life. In 1956 I returned once more to Pacific entomology and was then very glad to do so.

'Corran Garden' was never a profitable business; but we had established a satisfactory sales outlet, and our tomatoes possessed in quality what the crop lacked in abundance. After war broke out, additional capital became available from the will of a relative. This enabled us to build a good house on the site and put up more glasshouses.

When war began I was denied service with the armed forces on account of my occupation as a food grower. I joined the Royal Observer Corps, in a part-time capacity. Apart from the constant routine reporting of weather, movements of the RAF and shipping, there were few hostile aircraft to report. Two attempts were made to bomb the aluminium works at Fort William, both abortive.

I did have some occasional contact with entomologists. Anthony Downes, from Glasgow University, was studying the Highland midge problem. He sent me a copy

Jessie MacLean with a good Corran tomato crop

70

of his report. Later on, Guy Morrison, an entomologist at Aberdeen University, sent me a 'Pelican' paperback about biological control by Hugh Nicol, a Rothamsted chemist. In it were detailed accounts of Taylor's and my projects on Fiji's pests. That was a nice surprise and a considerate gift.

That little rascal *Culicoides impunctatus*, the worst of the midges, plagued us a lot at Corran; but we were greatly helped again by Morrison, who supplied us with a small bottle of dimethylphthalate (DIMP), a mosquito-repellent used by Australian and American troops in the Pacific war. We impregnated squares of camouflage netting with DIMP and wore these as veils to hang over our eyes. When not in use each worker kept the veil in a tightly lidded tin.

We had little trouble from *Pieris* (cabbage butterfly) on the brassica plants. Our worst enemies were nematodes attacking the roots of tomato plants in the glasshouses. We had in the end to resort to steam-sterilising the glasshouse soils; hard work and expensive, but effective.

PART II — 1956–1966

Return to the Pacific

After being a market gardener for twenty years I was glad to be able to return to Pacific entomology in 1956. During the previous year we had had visits from the Roths and the Garveys during their leave in England. These not only kept alive contacts with Fiji — sustained during the war by my occasional broadcasts in 'Calling the Islands' programs — but Garvey had by then become Governor of Fiji and had asked the Director of Agriculture whether he would like me back to work on any further pests. The banana scab moth *(Nacoleia octasema)* and the coconut flat moth *(Agonoxena argaula)* were giving some trouble, so an approach was made to the Colonial Office to prepare a scheme involving me as entomologist, to be funded mostly by the U.K. Government.

The project was duly arranged and I got a letter from B.A. O'Connor, then Government Entomologist in Fiji, giving me details about the pests and where he thought I should seek parasites for them, namely in New Guinea.

Although very glad to return to research in my profession, I felt extremely rusty and bucolic. There had been such far-reaching advances in entomological knowledge and techniques, both during and after the war, I felt I needed to get soaked in as much of the insect control atmosphere as was available, both in London and (en route to Fiji) in Canada, USA and Honolulu. I had discussions at the BM with John Bradley and Paul Whalley, specialists in the moth taxa with which I would be concerned. Talks were also had with Barnes at Rothamsted and Hall at the Commonwealth Institute of Entomology (CIE). The Crown Agents helped to equip me with apparatus, and P.K. Dutt Ltd supplied both instruments and some extremely useful materials. It was a great wrench leaving

wife and children, but I hoped Noel would be able to join me later.

I sailed from Liverpool on the *Empress of Scotland* in July. I felt this was preferable to flying. I wanted time to write and reflect on forthcoming plans and how they might best be carried out. There was an entomological conference taking place in Montreal, which Tothill was expected to attend. Unfortunately we missed each other. I went to see W.R. Thompson, who told me about the Belleville experiment station and put me in touch with G.P. Holland and E.G. Monroe, who were themselves expecting to visit New Guinea the following year. Monroe told me to be sure to meet J. Linsley Gressitt in Honolulu.

I went from Ottawa to the lab at Belleville. This large establishment had many applied entomologists among its staff, concerned mainly with Canada's forest pests.

They treated me, not as a tiro, but rather as a 'guru', one of a team which had produced a biological control classic. I felt at times rather an imposter and a bit embarrassed.

After Belleville I resorted to flying and arrived at Riverside, California, where there was a high-powered entomological research establishment at which many well-known American entomologists were

based. I was met there by Dr C.E. Lloyd of the CIBC, who took me to the Riverside station where Dr Paul deBach showed me round and introduced me to some of the top men working on citrus pests. These included Prof. Harry Smith (retired), and C.P. Clausen whose 'Entomophagous Insects' was to become my most used reference book during the next ten years. He was smoking a

Dr David C. Lloyd (CIBC) at the Fontana Laboratory, southern California, October 1956

huge cigar, was reputed to be the highest paid living entomologist and certainly looked like a man who had contributed much to solving some of the problems of humanity. Others I met were S.E. Flanders, C.A. Fleishman, H. Compere and T.W. Fisher. I felt I was on a fast lift (I should say elevator) and had ascended from the ground floor (at Ottawa) to something like the tenth, here at Riverside.

Harry Smith looked after me for several days, showing me American football and identifying the orange plantation once owned by one of my Paine uncles. Then the Lloyds put me on the flight to Honolulu. There I looked down from the balcony of my hotel room at the first coconut palm I had seen for over twenty years. I gave it a friendly wave. Beside it was a blue swimming pool, the whole scene like some holiday brochure. For me it was as soul-lifting as was seeing bare trees in England

Waikiki from the Princess Kaiulani Hotel, Honolulu, October 1956

in winter after four years in Fiji. Some folk seem contented to stay put all their lives in one place. This was not for me. I have an inbuilt urge to travel. During recent years at Corran I had felt incarcerated. The interest of starting and developing the market garden had worn off. It had become more and more a routine job beset with intractable problems.

I telephoned Dr Gressitt and introduced myself, and he said he would collect me and take me to his home. Mrs (Peg) Gressitt insisted on me staying with them and their four school-aged daughters. Peg Gressitt, besides helping her husband, was an accomplished pianist. Lin Gressitt was a man who had been studying Pacific insects, on land and high in the air. He

C.E. Pemberton (left) and J.L. Gressitt, Hawaiian Sugar Planters Association, Honolulu, October 1956

seemed quite the most devoted student of insect life I had ever met, and always able to get the funds to pursue his studies. He had his laboratory and made his headquarters at the B.P. Bishop Museum. I was to see much more of the Gressitts, in New Guinea, Hawaii and even at Corran, after I returned there in 1958.

The search for banana scab moth parasites in New Guinea

After my interesting talks with Gressitt I flew on to Fiji. There were many changes since 1934, including the international airport. Inconveniently, this is on the wrong side of Viti Levu for Suva, which remains the seat of Government; so I had to take a small aircraft to fly from Nandi to Nausori, the nearest airport to Suva. Barney O'Connor, with whom I was to collaborate for the next ten years, met me and drove me to Government House.

There Ronald and Pat Garvey gave me a warm welcome and asked me to stay with them. This was red-carpet treatment and I was shown to a bedroom the size of a badminton court.

Suva had grown: new roads, new suburbs, new shops. After three days at Government House I was invited to stay with George (Kingsley, as he was better known) and Jane Roth. Each day I was driven to Koronivia, about 16 miles from

Banana packing station on Wainibuka, Fiji

78

Suva, which was the new research station for the entomologists and chemists. O'Connor lived there. I met his wife Mary and their children. V.S. Pillai, assistant entomologist, drove me to see a banana-packing station, where I had my first view of scab moth damage. This was near Vunindawa, where at a nearby village there was appreciable damage by *Levuana* on some tall coconuts. There were only small larvae, but already *B. remota* eggs were on the largest of them, still only third instar.

Method of dusting bananas for scab moth control, Navuso, Fiji

O'Connor thought the search for scab moth parasites should begin in New Guinea. He had made brief visits to its mainland while he was working at Keravat (New Britain) and was struck by the absence of damage to bananas there compared with the situation in New Britain. He had bred a tachinid from *N. octasema* at Rabaul.

So to New Guinea I went. Before leaving Fiji I spent a few days on Taveuni. I was asked to check the effectiveness of metal bands on coconut trunks to prevent rats climbing up. Barney also wanted me to bring him a collection of eggs of *Graeffea,* the stick insect which caused much damage on certain estates.

En route to New Guinea, I made my first visit to the entomology department at the CSIRO, in Canberra, Australia. A.J. Nicholson was chief and D.F. Waterhouse his deputy. It was my first meeting with Doug Waterhouse, with whom I thereafter had many contacts, both at Canberra and in our Cambridge home, where he came to visit us in the early eighties.

Scab moth damage to banana bunches

I flew on to Brisbane, where Veitch (late of the CSO, Lautoka) told me scab moth occurred in north Queensland, but wasn't a serious pest.

The Suez political crisis loomed large in the news as I joined the steamer *Soochow* for Port Moresby. I cabled both O'Connor and Sir Ronald Garvey wondering if I ought to proceed. Then I heard that the Duke of Edinburgh had been in Port Moresby. He was due to open the Olympic Games in Melbourne. He also wondered if he should go on. He was told by his wife to do so. Based on such regal exhortations I decided to do the same. At Port Moresby I met Larry Dwyer (Chief) and Frank Henderson, his deputy at the Agriculture Department. The *Soochow* put in at Lae, where I was expected to work, but I continued on board to Rabaul so I could meet Gordon Dun, Senior Entomologist, Papua New Guinea, who resided at Keravat. Dun's wife (née

Bonnie Crabbe) had been my girlfriend in Suva at one time, so she received me with affection. When I left, I omitted to kiss her and she shed tears, which made me feel very guilty. The Duns put me up and there I had the first sight of scab moth as a serious pest. I soon learnt that the stain from immature banana leaves an indelible brown mark on clothing. It was difficult enough to remove from hands.

It seemed a bit odd that *N. octasema*, if it occurred on mainland New Guinea, didn't appear to feed on cultivated bananas at all. I found it difficult to believe that this was due solely to natural enemies. It seemed more likely that there were biological races of the moth with different food preferences. It was known that the larvae fed on *Pandanus* flowers in Fiji. Besides these, there were several *Heliconia* varieties, relatives of banana, which could also be host for scab moth.

At Lae, in the Morobe District of New Guinea, I made headquarters and contacted John Womersley, head of the Botany Department. He offered me space to work in his laboratory. Womersley did a lot of field collecting, knew the ropes about transport and was fluent in 'pidgin'.

I got larvae from *Heliconia* and *Pandanus* and left them with John to rear, while I visited Umboi (Rook Island) which lay in the Vitiaz Strait, half way between New Britain and New Guinea. While there I got a cable from John to say that the moth bred from the *Heliconia* larvae was *N. octasema*. It resembled closely the pinned specimens of the Fiji moth I had left with him. At that stage I was not sure that there were biological races of *N. octasema*, morphologically identical; except perhaps for their genitalia, which I hadn't examined in any detail. It was rather a tricky situation. O'Connor wanted me to work

at Lae, where any parasites would have to be collected from *Pandanus* or *Heliconia*. I returned to New Britain and collected a small number of banana-feeding larvae to take to Lae where I would rear adults and compare them with the New Guinea moths. Dun knew of my intention and although he didn't say I ought not to do this, he phoned Henderson who had me met off the flight and told me at once to take my collection to Bubia where it must be destroyed. At the time I felt annoyed and rather angry, but when the facts became evident I realised that Dun and Henderson had been justified in minimising the risk of releasing the banana-feeding race into mainland New Guinea.

I made a second visit to Umboi where I stayed in an isolated wooden shack belonging to the Lutheran Mission. After getting into bed I heard a rustling noise behind the shack. Taking a torch I went out to investigate. I withdrew in a rather cowardly manner when I saw an offensive looking animal with spines on its legs crawling about on the litter. I got it into my collecting net, then into a box with chloroform so I could show it to Keith Nagel, the missioner, in the morning. In the morning he told me the black stick insect, for that was what it was, was fairly common. The curved spines on its femora were used by the natives to pierce their ear lobes.

Nagel asked me to tell his school class the purpose of my visit. I gave a brief account in plain English, as I hadn't learnt to speak pidgin correctly. When I attended a service in his church, my work was the subject of Keith's sermon. It was told thus. 'Dispela binatang (insect) im i bagarup pisang (banana) alageta tumuch'. The phrase 'bugger up', used from the pulpit seemed odd, but in pidgin it merely means 'ruin' or 'destroy'.

Collecting scab moth in New Britain

There was definite, but limited, scab moth damage to bananas on Umboi, so it occurs within 40 miles of the mainland. To explore more of the island I visited a village called Gauru. I had with me an interpreter. Shortly after getting into my tent to avoid mosquitoes, he called to me: 'Master, master come quickly as the people are arguing and squabbling and accusing one man of not doing his job. If you come they will ask you to judge the matter and they will do what you say'. Although inexperienced in legal matters I hurriedly pulled on trousers and shirt and walked to the scene of the action. I gleaned that the accused man should have cleaned a latrine, as it was his turn. He disputed it, and he had a few supporters. Only guessing how to resolve the matter I pointed to the accused and told him to get on with the job. Immediately the fracas ceased. Judgment had been delivered, and everybody was happy.

When I had been in New Guinea a few months I realised it was essential to have a car. I bought a second-hand Holden utility with room for two or three in the cab, and a large open truck-like body behind. I had to persuade my Fiji HQ how necessary such a vehicle was to collecting quantities of the host plants from which it could be expected that *Nacoleia* enemies would be found. There was no provision for me to buy a car. It was an occasion when a decision should be left to the man in the field. After some pressure they relented in the end, and allowed the expenditure.

Womersley arranged for me to go with him and Professor Ronald Good (botanist from Hull University) to Nondugl in the highlands. There an experimental station had been set up for the study of wildlife. It was the brainchild of Sir

A family preparing sago, Umboi (Rook Island)

Edward Halstrom, who made a fortune selling refrigerators in Australia. He also financed a Zoo at Nondugl. It was 5400 ft up and pleasantly cool in the evenings.

At the commodious rest house we found ornithologist Peter Scott, his wife Philippa, and Charles Lagus, a BBC cameraman. We all sat round a log fire after the welcome evening meal, as our party had existed on a biscuit and two small sandwiches since early breakfast at Lae. I was beginning to realise the unpredictability of travel in New Guinea. Philippa Scott had a tropical ulcer on her leg. I knew how unpleasant that could be — before there was penicillin. I asked Peter if he had any ambitions to travel polar regions like his famous father. He said he had inhibitions about that. Instead, his mother had decided he should be trained as a zoologist, and his mentors were Julian Huxley and Kenneth Fisher. Peter was at Nondugl to study a rare species of duck. He had also organised a 'sing sing', which Lagus was to record. This was a splendid sight, which our party was lucky to see. I was glad to have a colour film in my camera and be able to get pictures of the elaborate costumes of the Papuan performers.

Our party then got in a Landrover to Banz, where, on the coffee plantation of a *Soochow* fellow passenger, I was told there were plenty of banana plants not far away. I looked them over. There was no sign of scab moth.

After returning to Lae, I began to collect *N. octasema* larvae from *Pandanus* flowers growing in fairly isolated groups up the Markham valley. From these I bred out the polyembryonic egg/larval parasite, later to be described by Eadie of the BM as

Fruit of Pandanus *in which scab moth pupae were found*

Pentalitomastix nacoleiae, an encyrtid. I felt inwardly excited to have found a parasite which attacked the only fully exposed stage of scab moth's development. I wondered if this would solve the scab moth problem for Fiji.

I began to hold a steady supply of *N. octasema* in the lab. Fertile females lay eggs on the side of 4 x 1 inch glass tubes. After exposure to *P. nacoleiae* thousands of the parasites could be bred easily and sent to Fiji. I must have communicated my enthusiasm to O'Connor, as the potential destruction of the scab moth pest appeared in the *Fiji Times* and was copied by an Australian journal. It was not long before I began to realise the folly of my ill-judged optimism. Firstly, there was a hyperparasite attacking some of the *P. nacoleiae* I had bred at Lae and I wasn't certain that this would have been excluded from the consignments sent to Fiji. I cabled Barney to make a careful check on this, as the

hyperparasite (*Tyndaricus clavatus* Eadie) could be distinguished by the shape of its antennae. It was certainly a very anxious moment. It was not too difficult to purify my breeding stock of *P. nacoleiae* and over the next few months thousands of this parasite were released in Fiji.

Although recovered once, quite locally, the species never established. The reason for this, as I feel sure is also the case for many parasites, is that they are attracted initially by the plant on which their host feeds rather than by the host itself. Had I realised this at the time it would have been evident that none of the New Guinea parasites from *Pandanus* or *Heliconia* would have been likely to be of use for the banana-feeding scab moth. It was certainly very disappointing.

Noel came to join me at Lae in February 1957. We rented a house in Coronation Avenue from the Seventh Day

Adventist Mission. It was only a few minutes walk from the Botany Laboratories and Botanic Gardens. It didn't appear to have a name so we called it 'Pandonia', which seemed appropriate.

Lae was renowned for earth tremors known as 'guria'. Occasionally there was a big one, as at 11.30 pm on 24 April. It may not have registered much on the Richter scale, but it got us from our beds and, staggering along the passage, we watched the water in the big open-air rain water tank moving first one way and then another. After the guria subsided it began to rain and five and a half inches fell during the 25th.

Bishop Strong, Primate of Papua New Guinea, came to preach in the small Anglican church at Easter. We had him to tea. He told us he started in a Leeds slum parish, then came to PNG where he had been for 20 years. Other visitors to Lae, in particular to the new herbarium and Botanic Gardens, were Robert Menzies, the Australian Prime Minister, his wife, Dame Pattie and his Native Affairs Minister, Paul Hasluck. They had come from Japan and looked far too hot in their grey suits.

Much more interesting visitors for us were David Attenborough and, for the second time, Charles Lagus. They were on their way to Nondugl to film and collect Birds of Paradise for a series of talks ('Zoo Quests') for BBC TV. They had tea with us and Attenborough talked continuously about his adventures. It was fascinating listening to this man who seemed to be blessed with such remarkable powers of communication (a capability I seemed so sadly to lack). After tea I offered to take him to the edge of the Botanic Gardens where I had seen the 'Victoria Regina' Bird of Paradise. It was in rather poor light, but the birds were there and I felt I could claim to have been the

Noel at 'Pandonia', Lae, 1957

first to show this bird to a man who eventually became one of the world's great naturalists.

We didn't see them again before they left PNG, but David wrote us a long letter describing how he got his animals back to London after a series of adventures; and the problems of temperature he had to overcome, taking animals from the cool highlands to the heat of

Rabaul. The whole journey back to England took him a month, instead of the twelve days he had allowed for it. He did well to get his birds and beasts back without loss.

To see how *N. octasema* from *Heliconia* would thrive on banana flowers I transferred larvae from the former to the latter. They all died after only a few nibbles, but did feed when confined in a breeding box.

Bay to the west of Busama Village, near Lae

Fruit bats (flying foxes) robbed us of bananas and soursops growing in our garden. The natives wrap dead banana leaves round ripening banana bunches to protect them from these pests. But they also eat the bats and use their bones to make strong 'net' bags (billams).

In August, the Canadian entomologists, George Holland and Gene Monroe, arrived at Lae: Holland to collect fleas of animals; Monroe to collect moths. They stayed at the Cecil Hotel, but used our house as their working centre. We lent them our Holden when we didn't want to use it. This must have been a great facility, for it would take their mercury vapour light to collecting sites in the evening. I found it fascinating to watch Monroe, recognising the species of many of the scores of different microlepidoptera attracted to the light. After an evening's collecting Gene would mount and label moths in the morning, putting them into special boxes which gradually filled the floor of our sitting room.

The Dutch entomologist Simon-Thomas from Hollandia (as it then was) called on us with a small collection he was taking to Szent-Ivany for identification. He left me a short paper on his local fauna, which was useful when I later worked on rhinoceros beetle.

Parasites for coconut flat moth

It was about September when O'Connor wrote to ask if I could study the little coconut flat moth, *Agonoxena argaula,* to see if there were parasites on it in Papua New Guinea. It was a tiresome little insect to rear as its eggs are laid along the midrib of coconut leaflets, which, shortly after they are cut, begin to curl up so the eggs get squashed. Also it was scarce and took much time collecting. There were several parasites attacking it, some of which I sent to Fiji. *Agonoxena* became of greater importance to us after we moved to Rabaul.

An amusing incident was the discovery, by John Ardley, entomologist at Bubia, of a large caterpillar which ate the leaves of a weed in cocoa groves. Unfortunately for him the natives took a fancy to the caterpillars, which they found good eating.

New Guinea was a mecca for visiting scientists. A few that

we encountered have already been mentioned. Others to visit us were Elmo Hardy (dipterist) from Hawaii; John Smart, collecting simuliids, from Cambridge; and Sedlacek — a refugee from Prague, collecting for an Australian museum. Our house seemed to attract them.

We went to Rabaul to probe the *Agonoxena* situation and found it more promising than at Lae. There was evidence of attack by a tachinid. At Keravat, A. Catley, a young Australian trainee entomologist, was in charge while Dun was on leave. He was always called 'Mick'. We became very friendly with Mick Catley, who, when I joined the Rhinoceros Beetle Control Project in 1965, was in charge of it.

There were reports of a severe earthquake at Lae, so we returned there in an anxious state, wondering how our livestock had fared. They were

unaffected and so were the low-watt lamps I had installed in cupboards to counteract the excessive humidity. The rainfall in 1957 was 238.5 inches (average 180).

Much of the *Heliconia* material we examined came from the Mission at Bumayang, a few miles from Lae, to which we were invited for their open day and bazaar. The Mission pupils provided interesting music: firstly on conch shells of varying sizes and then on bottles. These were of different shapes and sizes, part filled with water; the depth of water controlled the note when the boys blew across the top of them. With this novel kind of bottle orchestra they gave a recognisable rendering of 'From Greenland's Icy Mountains'. We much enjoyed this diversion and bought some nicely carved wooden tableware, and a pair of little salt and pepper 'people'.

Collecting Agonoxena *at Rabual*

Holland and Monroe had left to carry on their collecting at a hill station called Wantoat (alt. 3800 ft). They stayed with the District Officer, Colin deAth. They sent us glowing accounts of the place. I thought it worthwhile to find out what *Nacoleia* parasites there might be, in a cooler climate, more like Fiji's than was Lae's. We chartered a light aircraft to get there. The small landing ground at Wantoat was approached through a fairly narrow defile in the hills. It was an excellent chance to get a plan view of the Lae environs and the Markham valley, and to observe the extent of the soil eroded by its river which spread from the mouth in a large fan into the sea.

Colin was away on patrol when we arrived, but his house boy was fully up to the needs of unexpected visitors and soon had food and beds ready. Our letter to Colin was slower getting to him than we were.

A tall, hitherto unknown *Heliconia* with red stems was plentiful there. From it we obtained *P. nacoleiae*, showing this small wasp to be widely distributed. One never knew what excitements might present themselves in New Guinea.

In early January 1958, I visited Rabaul again and collected a quantity of *Agonoxena* parasitised by a small tachinid. This was later described by Crosskey as *Actia painei* sp.n. So far as I was concerned it was aptly named. Of all the tachinids I have attempted to breed, this was the most intractable. It was an odd little beast. The adults, as soon as they emerged, flew up and battered themselves to pieces against the top of the cage. I was so baffled by this behaviour I could send to Fiji only parasitised hosts and hope that the abundance there of *A. argaula* might enable the parasite to get established. It was a vain hope, I'm afraid,

because the moth hosts in New Britain were never plentiful and in mainland PNG even less so. *A. painei* did turn up on Bougainville, near Kieta, but never in such abundance that one could collect more than a dozen or two at a time. I felt that if anyone could devise a way to breed this tachinid it might do good work in Fiji, as being less likely to suffer from attack by hyperparasites than parasite species earlier introduced.

Returning to Lae later in January, we met Harold St John, Professor of Botany in Hawaii, one of the few specialists on *Pandanus*. He thought there might be 100 species of these in PNG, as he had described 130 species in the Philippines. Besides being a host for *N. octasema* on its flowers, a tall species was fed on by the zygaenid, *L. gracilis* in the highlands.

Fiji and home leave, 1958

We packed up and left Lae early in February 1958. I had been there 16 months and Noel 12, so we felt well acquainted with the place and were quite sorry to leave. Once more at Rabaul, we made headquarters at the Department of Agriculture. After searching the area for *Agonoxena* we found a place at Talvat, at the back of Matupit volcano, where we could collect 250 larvae from 20 coconut leaves. There was a good percentage parasitisation by *Actia* and it was mostly from that locality that this parasite was sent to Fiji. Before my first contract for the scab moth work concluded at the end of February, we collected about 20 of the tachinid parasites O'Connor had wanted. That was all we could send him after two weeks collecting.

The family at Corran, 1958

We sailed from Rabaul on the *Malaita*. Its skipper was the well known Australian sculptor and painter, Brett Hilder. During the trip to Sydney a young nurse fell from her bunk and had her leg put in plaster. Brett used her leg as a canvas for one of his paintings.

From Sydney we sailed for Suva on the Orient Line *Himalaya*, and had a few days of luxury after the rough and tumble of travel in New Guinea. In Suva I had talks with O'Connor and Watson, the new Director of Agriculture, and also with the Governor. They were anxious that research on the banana scab moth should continue; the coconut stick insect *Graeffea crouanii* should also be studied. I suggested a further two-year project, with searches in Malaya, Java and Queensland, involving a grant of fifteen thousand pounds, most of it to be met by the British Government.

In January 1959 my family and I moved from our West Highland home, having sold its market-garden, and moved to the Old Vicarage, Syleham, in north Suffolk. That was to be our home address until 1975.

A new contract — work in Singapore, Malaya, and Indonesia

My new contract began on 1 April 1959. First I went to Holland to have talks with some of my former colleagues at the Bogor Institute. The microlepidopterist had worked in Celebes (Sulawesi) for some years and saw no sign of scab moth on bananas there.

I bought a single-lens-reflex camera and field glasses with a minimum focus of 25 feet. The Bishop of Singapore, a family friend, invited me to stay with him and introduced me to Humphrey Birkhill, Curator of the Botanic Gardens. Here I made a careful study of the palms and pandans. The name 'pandanus' derives from the Malay word denoting fragrances. This was apt. I found I could detect a flowering pandan at some distance by its scent.

I collected *N. octasema* larvae from the flowers of the *Nipa* palm, growing only a foot or two above sea level. I took them to Kuala Lumpur, where they fed normally on pandan flowers. Ahmad bin Yunus, the young entomologist there, became a good friend and came on several field trips with me later.

Back at Singapore I found much political furore, leading to the election of the People's Action Party under Lee Kuan Yew, a Cambridge graduate, as Prime Minister, a position he held until 1990.

I flew across to Sumatra, where, neither around Medan, nor at Brastagi in the hills, was there any sign of scab moth damage to bananas. My next search was to be in Java. I was invited there by 'Prof.' Kusnoto who was now in charge of the Kebon Raja.

In Java, scab moth was everywhere evident on bananas. Here, in fact, was a parallel situation to that between New Guinea and New Britain: Java and New Britain affected; New Guinea and

Sumatra unaffected. In Java *N. octasema* was also present on pandan flowers and on *Nipa* in the Botanic Gardens; but not on *Heliconia,* a genus not indigenous to Java. I wondered how far eastward bananas would be infected, so joined an expedition to Bali where it was evident on bananas up to an altitude of about 3000 ft.

I had also to take note of *Agonoxena* and found it had a closely related genus *Haemolytis* feeding on coconut and *Metroxylon* (sago palm). A young entomologist at Bogor, called Sutyoso, spoke good English and seemed only too glad to help me with my project.

On an expedition to a small island off the north coast, I set down my haversack while I walked about examining various plants. Returning to retrieve my bag I found it wasn't there. It had been stolen. Similarly, on the streets of Jakarta I was often jostled and, on one occasion, found my pocket pen had been taken.

One compensation amid the interminably dull meals provided at the Kebon Raja Rest House, was the durian. The unsociable odour of excrement which this fruit has is a deterrent to many people, and makes it unacceptable in some of the higher class hotels; but for those, like me, who appreciate their unique flavour, they are, as A.R. Wallace records in his 'Malay Archipelago', superior to all other fruit.

An opportunity came in September to discover the extent of banana-attacking *N. octasema* in islands east of Bali. The marine research vessel *Samudera* was to sail

Noel and 'assistants' at Panimbungan, Java, after collecting Haemolytis, 1959.

from Java to Ambon, calling at Banjarmasin and Manado en route. Movements ashore were restricted by the military, who ruled the roost in this period of confrontation with Malaya.

On the ship I had quite a lot of apparatus with me, including a binocular microscope. All this was stowed below deck; but so were dozens of sleeping bodies, wanting to visit places away from Java and taking any chance to get to them. I was at any rate allotted my own cabin.

I could see no sign of scab moth at Banjarmasim and Manado. This made me wonder how it had got into Java. After a week at sea we reached Ambon, where I met Dr Wegner, who was head of the Zoological Museum in Bogor. He had heard about my search for banana scab moth and told me it wasn't on Ambon or Buru Island and he had never heard of it outside Java.

The laboratory at B.P.H.I., Bogor and a Haemolytis *breeding cage*

95

I had ten days to spend at Ambon — time to survey much of it and get to know many of these pleasant and friendly Moluccan folk.

For the last five weeks there had been no aircraft reaching Ambon owing to lack of fuel. A tanker was overdue. I felt quite out of the world and had had no letter from Noel for four weeks. She, on her part, was so concerned about my welfare she rang the Colonial Office, which contacted the Consulate in Jakarta. They said as far as they knew I was all right, just temporarily out of touch.

On 2 October I had to transfer from *Samudera* to the Provincial ship *Babut*, as the *Samudera* had to return with Wegner to Java.

The *Babut* was scheduled to sail on 12 October. During the ten days that this rather squalid little vessel was tied up at the Ambon wharf I was nearly roasted alive. One evening I was visited by a rather elderly prostitute who spoke enough English to convey that she was quite used to elderly tuans. I told her I was married and didn't require her services, at which she gave a rather hollow laugh and drifted away. I was nevertheless relieved that she wasn't as young and attractive as some Moluccan girls are!

The only interest I had during that gruelling period was joining an expedition tuna fishing. That was quite interesting and, as the fish were hooked with bamboo fishing rods from the deck of the launch, I was reminded of evening 'cuddy' fishing in the narrows at Corran, except that tuna are about ten times the weight of a saithe (cuddy).

The last of the *Haemolytis* larvae I had brought from Java died. Coconut leaflets stay fresh for only two days and each time fresh food is provided a larva uses much energy spinning the web beneath which it always feeds.

On 13 October the *Babut* sailed for Banda. On that small spice island there were signs of scab moth on bananas, although minimal. I left a local friend there to collect as many larvae as possible which I could get on the ship's return. I continued on the round of this inner arc of volcanic islands, concluding at Roma, many of them mere hills of bare rock and with no further signs of scab moth.

On 18 October the *Babut* returned to Ambon. Efforts to get scab moth larvae from Banda were unsuccessful. Like much else on this phase of my travels I was hampered by lack of the local vernacular: my limited (pasar) Malay didn't seem to convey my meaning, nor could I fully understand what was said to me. It was frustrating.

I had to hang about for nine more days at Ambon. The only flights to Java were booked up by army personnel. When I did eventually get back to Java I reported my return at the British Embassy.

I was anxious to conclude the survey of the islands to the east and flew on November 7 to West Timor. There was no hotel accommodation available, but the Director of Agriculture, Mr Pelapelapan, although he had five small children, made available a room in his house. He and many others in these outlying islands were Christians, very kind, and could speak English. There was the usual round of official visits — Army, Police, Governor and Immigration. To avoid any delay when departing there could be no unrecorded entry.

On Timor there was ample *N. octasema* damage to bananas. I had found it absent on the smaller islands to the east so the banana-feeding race must have spread through the lesser Sunda Islands from Java, and therefore could be expected on Flores. On November 12 I flew from Kupang to Maumere on Flores. The whole place looked very parched. I went to stay at the Roman Catholic seminary at Ledalero at an altitude of 650 ft. Father Baack, a Dutch priest responsible for the agricultural production of the seminary, took charge of me and drove me about on the pillion seat of his motorcycle.

Young insect collectors near Tjilogon, Java, 1959

After nine days collecting scab moth from bananas a *Chelonus* was bred from them. As this parasite oviposits in the eggs of scab moth it should be an ideal parasite, but this genus was not in general considered very effective; the species usually seem to lack the ability to discover all available host individuals. If they had the searching ability of *Apanteles* they would be extremely valuable. Some genetic engineering seemed needed. At Ledalero the *Chelonus* was achieving about 5% parasitisation. Hoping that this Flores *Chelonus* might prove the exception to the rule I took enough back to Bogor to start a breeding stock. It was an easy parasite to breed, since, confined in a tube with scab moth eggs, it could hardly help finding them. Leaving the routine breeding to Sutyoso I flew to Singapore to collect Noel.

During 1959 I had made 19 flights, two steamer journeys and one bus convoy. I felt

much travelled and had been without my wife for nine months.

Back in Bogor we were accommodated at the Kebon Raja Guest House, but our laboratory was at the old Institute. Sutyoso, excited about the scab moth parasite, had already released it near Bogor. He showed me a garden with many varieties of banana. In the market one can find 13 varieties, of which the 'pisang Ambon' was considered the best dessert banana.

On 19 January 1960 we sent the first consignment of *Chelonus* to Fiji. We had to make sure that the parcels of parasites were kept in the cabin; it would have been too cold for them in the hold of an aircraft.

Haemolytis larvae had been difficult to collect for the previous few months, but with the rains beginning they became more plentiful. A

Macrocentrus ichneumonid was attacking the larvae and we bred enough of it to send it to Fiji. Banana scab moth was also more plentiful in the wetter weather. In a kampong about 500 ft higher than Bogor we found a banana plant with 1227 eggs on it in 58 clusters. During this time I saw, for the first time, adult scab moth in the field. By beating the dead foliage of a banana they can sometimes be flushed into brief flight. They match the colour of dead leaves so exactly they are easily overlooked.

On 13 April we flew from Java to Sydney, then by other flights to Cairns for the study of scab moth in north Queensland. We centred at Innisfail and made contact with George Ettershank, the local entomologist. We looked at several banana plantations with him and Mr Lovelady (what interesting names some of those N. Queensland folk had). In that land of 'do it yourself', I carried an axe in the car, and felled a tall, slender 'cabbage palm', later determined as *Archontophoenix alexandrae*. It was exciting to find agonoxenid larvae feeding on it. This was a useful discovery which I took advantage of in a later visit.

Felling a 'cabbage palm' at Tully, north Queensland, April 1961

We went as far north as Port Douglas, and then returned to Brisbane. We bred only two tachinids from the banana scab moth collected. This was an introductory survey as we were due to take up residence again at Rabaul. There we were fortunate to rent two adjacent flats in a newly opened block. It was a productive time, when several consignments of the tachinid, *Argyrophylax proclinata* Crosskey (O'Connor's '*Bactramyia*'), were sent to Fiji. We found a place where a little *Actia* was reaching 60% parasitisation on *Agonoxena*. Five hundred of them were sent to Fiji, but it failed to establish. As already mentioned, I was unable to devise a way to breed it.

We were visited in Rabaul by John Womersley, who was conducting the British botanist Professor John Corner in a search for varieties of figs. At the Buka Island agricultural station the scab moth tachinid *Sisyropa painei* Mesnil was

fairly common. It ranged through the Solomons. Over 300 puparia were sent to Fiji, but not enough to get it established.

We wanted to get to Honiara, but were deterred at first by Mr Spencer, the Director of Agriculture, who said there was no accommodation. Not put off by that pessimistic attitude, which had proved unjustified on some previous occasions, we just went to Honiara. We found easy accommodation at the Mendana Hotel, thanks to the help of David Meadows, much less of a pessimist than his boss.

On 22 November Noel flew back to England, arriving to find our new home cold and messy, after a let in our absence; it was a bit tough on Noel, I'm afraid.

I sailed to the western Solomons on the *Coral Queen*, a government vessel. Banana plants were scarce at

Munda, so I couldn't get much scab moth; but I did notice tachinid eggs on a large stick insect, so concluded that parasites for *Graeffea* might be found in the Solomons.

From Honiara I got in touch with Father Wall, a Roman Catholic priest. He had local friends on the small volcanic island, Savo, where I went to spend Christmas. Coconut-feeding stick insects are more abundant on Savo than anywhere else in the group and as they seemed to lack attack by tachinids it suggested the latter might be controlling the pest on other islands.

Early in the new year of 1961, I flew to Rabaul to get some of my equipment sent to Port Moresby; the rest to Fiji. I continued in an old Dakota to Manus (without scab moth) and thence to Port Moresby. Next I flew to Cairns, where, at the end of the dry season, scab moth is almost non-existent on bananas. When the rains

begin it becomes plentiful again.

I made my centre at a pub at Tully, where the landlord gave me the use of a small cubicle almost surrounded by beer casks. It made a makeshift lab. I began to cut down 'northern bangalo' palms *(Archontophoenix)* and from larvae of *Agonoxena (phoenicia* Bradley, sp.n.) I reared an *Elachertus* and a *Chelonus* and sent them to Fiji. Parasitisation by the latter ranged around 17%. I hired a Holden pick-up at Cairns in which I covered 3,300 miles during my nine weeks stay in the north of Queensland, with collecting sites so widely separated.

On 23 March I was back in Suva and discussing with Barney O'Connor the proposed stick-insect project. I also visited the place where the Java-bred *Chelonus* was released. It was established, but the degree of scab moth control was not sufficient to

obviate the need for insecticide treatment to obtain totally clean bananas for export. This seemed to make my project on scab moth completely illogical and a waste of public money. I felt frustrated and disappointed. I arrived back in London by a Qantas flight on April 8th, 1961.

I commuted, mid week, from Syleham to the Natural History Museum writing my reports for the scab moth and flat moth work. This continued until early in 1962. On February 20th Taggers took me to the Verrall supper where I talked, among others, to George Varley, Professor of Entomology at Oxford. There also was N.D. Riley, who had identified much of the material collected by Margaret Fountaine. Although she was distantly related to me by marriage, I had never met her, but read the books published about her diaries.

The *Graeffea* project

In mid-October 1962 I began work on the *Graeffea* project, using E.O. Pearson's room at the Commonwealth Institute of Entomology (CIE) located at the British Museum. Looking through the museum collection of *Ophicrania leveri* (the Solomon Islands close relative of *Graeffea*) I found a specimen, collected by E.S. Brown, on a leg of which remained the egg of a tachinid. I had museum photographs made, through the good offices of David Ragge, head of the Orthoptera collections. They were very useful when I wanted to show people in the Solomons what I was anxious to collect.

On November 5th I took a Qantas flight to Washington D.C. There I met again A.C. Smith who, as a young botanist, had stayed with us on Taveuni, and was now Director of the U.S. National Museum. I also talked to Gurney who collected *Ophicrania* on Guadalcanal during the war.

At Honolulu I stayed with the Gressitts. At the B.P Bishop Museum I talked to Miss Nakata, who had been working on phasmids. I had hoped that, in Hawaii anyway, there was someone studying stick insects, for this seemed to be a very neglected group of insects by most taxonomists. Sad to say, however, Nakata died a few months later after publishing very little.

At Suva, I dictated my scab moth paper for typing. Old H.W. Simmonds, now 85, came with me for a walk in the Tamavua bush. Shortly after, I flew to Savusavu with J.S. Pillai. At Tuvamila plantation the foliage of the coconuts was skeletonised by the feeding of *Graeffea*. I had never seen such damage by this pest. On Taveuni, what had been a lengthy outbreak of the pest was now almost finished. The insects feed also on some species of *Pandanus* and the fan palm *Pritchardia*. I walked to the crater lake with Dex Hinckley and saw a male

stick insect on the bush palm *Exorrhisa*. The damage done by *Graeffea* over Taveuni as a whole was probably less than 5%: only locally does its population build up to pest status.

From Suva I visited Navua, where, in a fairly extensive growth of the local sago palm *(Metroxylon vitiense)*, there was some stick insect feeding.

Proceeding to Solomon Islands the flight took me via Port Moresby and Rabaul. At neither place was there any palm or pandan damage by phasmids.

At Honiara, entomologists John Greenslade and Mike McQuillan met me. After settling in I was introduced to T.C. Whitmore, a Cambridge graduate and Forest Botanist

there. To my delight Whitmore showed an interest in the small palms growing in the shrubby forest behind the town. These were plants on which stick insects were feeding.

I visited Savo Island again and found the phasmid *Ophicrania leveri* as abundant as it was two years ago, with no sign of tachinid eggs on them. Father Wall of Visale Catholic Mission told me of a lad called Samueli from a Savo village. He was intelligent, so I signed him on. He did a good job for us.

On Guadalcanal *O. leveri* feeds mainly on *Metroxylon*. A few had tachinid eggs on them, but the groves of these palms were at Visale, a 20-mile drive from Honiara, so the transport problem loomed large. Moreover, not many of the stick insects there had tachinid eggs on them and none seemed to have maturing parasite larvae in them. One of the BM specimens had been

Coconut palms on Vanua Levu, Fiji defoliated by Graeffea

Savo Island, Solomons

I had small carrying cages made which I gave out to a few native collectors at Visale. They could load up to two dozen stick insects in each, so when I could get out there there was a reasonable quantity of material available for study after taking it back to Honiara.

It was only after I had visited Kolombangara in the west of the group that our main

collected at Suta, a place in the foothills of Popomanasiu, the highest mountain (8000 ft) on the island. Greenslade and I did an expedition there. It was very tough going and we had some excitement with rapidly flooding creeks, which sometimes separated us for a few hours until they subsided. After camping in the bush near Suta I found few signs of stick insects, so the five-day excursion provided little more for me than good exercise.

On 19 February 1963, Noel arrived from U.K. to take on the job of Project Assistant. I began collecting *Sisyropa painei* from scab moth larvae at Betikama. A few of these tachinids mated in cages, and although one or two females had eggs in the uterus, the method of infecting hosts, so successful with *Tirathaba*, could not be used for *N. octasema*. It was, however, a sideline worth studying.

Phasmid storage cage

collecting ground of *O. leveri* in sago palm swamps was discovered. The collection of these insects must be done at night when they emerge from daytime concealment to feed. I took Sam with me and in a swamp near Kusi we collected stick insects from small palms we could reach and 15% of them bore eggs of the tachinid. This seemed quite the best place to get supplies of the parasite (described by Crosskey as *Mycteromyella laetifica*). The problem was to get them to Honiara, where we could attempt to breed them, and make dispatches to Fiji. They had first to go by canoe through occasionally rough water to Gizo. From there to Munda airstrip by launch, then on a flight to Honiara. Casualties in transit were often 25%. They were unavoidable.

We hoped to get the Kusi tachinid established on Savo, so about 30% of the collections were released there. It didn't produce the result we hoped for and it seemed there was a possibility that to pupate successfully the fly needed a moist soil covered with a thick layer of leaf litter, such as occurs in the sago swamps at Kusi. On Savo, hosts were all on coconut, not sago palms, with a dry porous soil beneath.

At this stage I decided to go to Fiji, tell O'Connor how the work was going and see if he had any ideas for overcoming transportation problems. This resulted in our getting a 'cannibalised' Landrover. It could be sent, together with a Hereford-cross bull, on the *Coral Queen*, which had taken Solomon Islands athletes to the South Pacific Games.

The development of the fly had some interesting adaptations to cope with the

Ophicrania leveri *with emergence hole of tachinid species near base of leg*

attenuated body of its host. If an egg is laid on a leg of the stick insect the newly-hatched maggot has a long journey to get to its host's body, where it can feed. To achieve this it has lateral ambulatory appendages. When ready to pupate the fully grown maggot may have to drop 30 or 40 feet from a tall palm. Its cuticle and spiracles are heavily chitinised in consequence. After hitting the ground the maggot crawls into moist surface litter to pupate. The only part of the host into which the parasite is unable to gain entry is the wings. As the *Graeffea* wings are shorter than those of *O. leveri* there would be less likelihood of the parasite failing to enter its host.

We sent 959 fly puparia to Fiji over a period of four-and-a-half months. O'Connor started to breed the parasite at Koronivia. It attacked *Graeffea* successfully and a generation was reared on that host, but later on he was unable to maintain a breeding stock, so

Tachinid release site in Fiji

all remaining puparia received from the Solomons were released directly on Taveuni. Sadly, that did not result in establishment.

In March 1964 my contract had terminated and I made arrangements for Fiji's Assistant Entomologist, Satya Singh, to go to the Solomons for further supplies of the tachinid. He found parasitisation at Kusi much

reduced, presumably owing to all our previous collecting, and had no success.

In searching for *O. leveri* on small bush palms near Honiara, I was often deceived, not in mistaking the stick insects for foliage, but in mistaking foliage for stick insects. Resemblance between them was so exact.

During my short leave, the XIIth International Congress of Entomology took place in London and I was able to have talks with many of the entomologists I had met abroad. I also discussed with E.O. Pearson an offer of a further contract to survey the rhinoceros beetle (*Oryctes rhinoceros*) in Southeast Asia for the South Pacific Commission (SPC) Rhinoceros Beetle Project. This appeared likely to be an interesting assignment in a part of the world I was well acquainted with. I accepted willingly.

The Rhinoceros Beetle Project

The 11-month contract with the Rhinoceros Beetle Project began on 18 June 1965. I spent ten days at the BM studying dynastine beetles, elaterid predators and reduviids, assassin bugs.

I flew direct to Honolulu, where I talked to Lin Gressitt, who had done some work on *Oryctes* species. Then on to Pago Pago, where I had to change planes to get to Apia. Here was the headquarters of the project and where Chuck Hoyt and Mick Catley met me. Hoyt, who was head of the Project, showed me the early stages of *Oryctes rhinoceros*.

Rhinoceros beetle, unlike the other coconut pests I have worked on, does damage as an adult. The grub-like larvae feed on rotting wood, heaps of sawdust and such-like valueless substances. The adults bore into the crown of coconuts and some other palms to feed on the sweet embryonic leaf tissue. If, in doing so, they damage the growing bud of the palm it will die. This was rather exceptional, but by chewing at the unformed leaves the latter open with scissor-like cuts through the leaflets to produce the tell-tale type of damage characteristic of this pest.

Oryctes rhinoceros *grubs, pupae, and adult from oil palm logs, Kulai Estate, Johore, Malaysia, 1965*

Oryctes rhinoceros *boring its way into the heart of a coconut palm frond*

My assignment for this rhinoceros beetle work involved me in more individual direction, with a degree of protocol attached. I was given letters to influential people and told the order in which they were to be seen. It was all very different from my work for Fiji and I found it quite exciting. The excitement was mainly due to Hoyt's discovery of a certain species of nematode in the sex organs of male and female *Oryctes centaurus* on sago palms in New Guinea. These most interesting nematodes did not occur in *O. rhinoceros.* This seemed to indicate that if I could discover a place where the latter contained this sexually confined thread worm it might be the place where the pest originated, and hopefully where it might be under control.

There was nothing in the first place to show that beetles with these nematodes were any less fecund that those which lacked them; but the nematodes are present in such dense populations, almost certainly feeding on the sperm fluid of the male beetles, I felt they must be reducing their fecundity. In the end, it became clear that I had reached a wrong conclusion; but this was not evident at the beginning of my survey and it really became the nucleus around which my search was based.

Before I left Apia I had a rather strange ride to the other side of the island with a young German physiologist, Karl Marschal. He lay, holding a shot gun, on the bonnet of the Landrover and when a landrail attempted to cross the road he fired at it. If hit, the poor bird was then collected, and Karl would remark that we should have a very good supper. He duly cooked one bird and assured me it was a great delicacy. For me, it was impossible to eat, far too tough. I thought he must have teeth like a mongoose.

Mick Catley, holding Oryctes *from sago palm, Apia, Western Samoa*

Karl Marschal was acquiring the local vernacular and behaving more and more like a Samoan. He was destined to play an important part in the ultimate means by which rhinoceros beetle was controlled.

On 22 July, I flew to Australia, stopping in Fiji briefly for a chat with O'Connor and old Simmonds. The latter was still employed part-time breeding scoliid wasps from Madagascar. These attack rhinoceros beetle grubs and

Central leaves from coconut palm with extensive damage by Oryctes centaurus at Daru Island, Papua New Guinea

dynastines. *O. rhinoceros* had not been recorded in Australia.

I had an unexpected talk to Roger Crosskey in Sydney. He had described many of the tachinids I had sent in previous years to the BM, and he was just concluding his collections from New Guinea to round off a paper on the family in Melanesia.

My flight went via Perth to Singapore, where Humphrey Birkhill met me and offered me a bed. Next day he introduced me to the Botanic Gardens palm expert, F.X. Furtado, a Goanese, who had worked with Baccardi. This was interesting as we had met Baccardi's son, Nelo, in Scotland.

I regarded Singapore as my logical base, and spent some time studying the palms in the Botanical Garden, several of which had obvious damage by rhinoceros beetle. But I was due to search in Ceylon and India before the Malay

Simmonds seemed to think they would control the beetle, although they have a very minor role in its control. At Canberra Phil Carne gave me reprints on Australian

Peninsula. I landed at Colombo where I had a letter to Mr Felix Kirthisinghe at the Coconut Research Institute, Lunuwila. There, rhinoceros beetle grubs were plentiful in coconut logs. I dissected eight adults in which there were no sex nematodes. The method was first to kill the beetles with an injection of formaldehyde using a disposable hypodermic syringe, then dissect out the rather complicated sex organs.

Kirthisinghe had a car called a Borgward 'Isobella' — a somewhat ancient vehicle with a maximum speed of 25 m.p.h. We set out on the gentle climb to Kandy, but had to stop on occasions to allow the car to cool down. But it gave me a chance to discover that Felix was Catholic and married with two children. He spent a year at Rothamsted and was more of an anglophile than others I met in Ceylon, which was becoming gradually nationalised.

There were ceremonial elephants padding along the Kandy streets; in their droppings rhinoceros beetle larvae are said to feed. We visited the famous Botanic Gardens of Peradeniya where I took photos of the *Corypha* (Talipot) palm avenue. This palm was to become more important in later searches. At Kalpitiya in the north of the island, only a short distance from India, I got more beetles for dissection in the Lunuwila lab.

Leaving this material for my return I flew to Madras, and thence to Bangalore. There, at 3500 ft on the Deccan plateau, Dr V.P. Rao, CIBC entomologist, met me. He had quite extensive laboratory facilities for breeding many different parasites, for which there was a wide demand. All the work on rhinoceros beetle was done at Kayankulam in Kerala Province. I flew to Cochin to be driven in a jeep by Chandy Kurian to Ernakulam. Dr Kurian was an

Anglican Christian, although most of Kerala was Communist.

At Kayankulam I stayed at the rest house. Dr Lal, head of the Station, was a Hindu; a nice man, who, with his very beautiful wife, looked after me well. The laboratories at Kayankulam were of high quality, quite a contrast to the rather basic layout at Lunuwila. In the few rhinoceros beetles I collected there were no sex nematodes. In Kerala, coconut husks are garnered for the making of coir. After they have been wetted in pools, they are beaten and the fibre spun to make rope. I was fascinated to watch the whole performance.

I also watched three men digging over an empty paddy field with their long handled chop hoes (changkols). The speed and perfect synchronising of their actions was most impressive; they worked continuously — the sweat clothing their faces.

I flew back to Bangalore where Rao took me to examine date palms, said by Venkatraman to be preferred to coconuts by rhinoceros beetle. We found no evidence of this.

Back at Colombo, Felix K. took me to a fibre mill where I paid the owner to fell two badly damaged coconut palms. The fronds of these were removed and laid out in their age order. Every one had rhinoceros beetle damage and five live adults were removed. In spite of this continual damage, the palm, over the previous two years, had produced 14 mature nuts. Other less damaged coconuts nearby had an average of 26 mature nuts. Thus, rhinoceros beetle could have reduced yield by as much as 50%. There was nearly always much damage to palms adjacent to fibre mills, but it was difficult to discover exactly where their grubs were breeding. No sex nematodes were found in Ceylon.

Oryctes oviposition chamber in coconut stump

I had to make calls at Kuala Lumpur to clear the way for work in Malaya. The first was on D. Bickerstaff of the U.N. Technical Assistance Board. He said he would cable New York to arrange a Landrover for me in Sabah. Then I called on Dr Sedky of FAO. He acted as my anchor man, to receive mail and clear my work in Malaysia.

In Singapore I hired a car and driver from the local Automobile Association and was driven to Batu Pahat. There was lots of rhinoceros beetle damage to coconuts in Malaya and I thought it unlikely there would be any good controlling agents. I watched large blue scoliid wasps flying about and had a few collected, but I knew Hoyt had these already. I went to the Oil Palm Research Station at Chemara to meet Brian Wood the entomologist. He found that if rotting logs were covered by weed growth it concealed them and prevented beetle oviposition. On a walk with Wood I spotted a dead pandan, cut it down and found a grub of the native *Oryctes gnu*. In later collections of *O. gnu* from pandans I saw two pupae which looked diseased. As diseases were conditions I knew very little about I did no more than note the occurrence. A year later I wondered if I had been looking at the virus collected by A.N. Huger, who introduced it to Samoa where it was bred and used with much success for rhinoceros beetle control, both there, and later in Fiji.

Proceeding northwards up the west coast I saw labourers at Jenderata Estate with barbed hooks pulling rhinoceros beetle grubs from oil palm logs. It was a crude method of attempted control which did little to reduce the numbers of the pest.

At Kuala Lumpur, Yunus arranged for his assistant Majid to come with me to

Mr Nasir (Agriculture Officer, Temerloh, Malaysia) shows Oryctes gnu *to locals, 1965*

Kuantan on the east coast, a drive of 170 miles. I used the splendid rest house there as base for work along the east coast and centre of the peninsula. Inland at Temerlun on the Pahang River *O. gnu* was everywhere abundant. This large endemic species usually contains the sex nematodes. I had the local State Agriculture Officer as field assistant and interpreter.

At 1.30 one morning there was a heavy thunderstorm: lights failed and there was a flood of rain. Then the frogs piped up, with a delightful degree of harmony.

From Kuantan northwards 137 miles to Kuala Trengganu there was an extensive belt of coconut palms on an old raised beach of sandy soil. They had little sustenance and there were many dead trunks. After felling several I found *O. gnu* and *O. rhinoceros* in the same stem, but even there the latter had no sex nematodes.

Awang, my driver, often had a nap in the car while I was working. I approved of this as he did a lot of driving. On the road he had to avoid sheep, goats, dogs, cattle and at times even buffalo. Rhinoceros beetle was present on the fan palm, *Livistona*, a new host record. But I found no insect-controlling agents on adult beetles.

Towards the end of November I went northwards with Majid and Liu from Kuala Lumpur to visit Langkawi Island near the Thai frontier. *O. gnu* was present on the islands and there were 'Ibu' palms *(Corypha elata)*, which I was to have more to do with in Sabah. At the rest house on the island there was a young coconut palm bearing the notice 'Planted by the Duke of Edinburgh, 1956'. He certainly visited the outposts of the former British Empire.

The return journey to Kuala Lumpur had its misfortunes. At Ipoh, Awang got diesel put in the car by mistake. Later on I was axing down a tall coconut, which disobeyed my usually successful axe-cut procedure by falling towards the main road. The top of it hit the telegraph wires on the way down. These broke the top off the dead palm stem, which then fell on the road scattering rhinoceros beetle grubs in all directions. I felt pretty small at that moment and was saved embarrassment only by clearing away the rubbish before the arrival of any traffic. Thus ended the first part of my mission to Southeast Asia.

Included in my schedule was the island of Gan in the Maldives. After returning to Singapore I gave up my car and driver at Changi airport and boarded an RAF 'Comet' for the flight to Gan. It cost £100 return. I sat next to Flight Lieutenant Bangay — a journalist photographer who had contacts with the U.K. popular press. He wanted an account of my work, but in the terms of my appointment I was forbidden media publicity, unless cleared by Project HQ, so couldn't oblige.

Feeding tunnels of Oryctes rhinoceros *on a coconut palm trunk, Gan, Maldives*

Maldivians embarking on dhoni at Gan

There was intra but no inter-mating, so no chance of getting sex nematodes from *gnu* into *rhinoceros*. I said thanks and goodbye to Ahmad Yunus and took the overnight train back to Singapore. On 15 December I flew to Jesselton (Kota Kinabalu) in Sabah. There Tay Eon Beok met me. He was taking on the entomologist's job from Gordon Conway. We went to Tuaran, the Sabah Agricultural Station.

There was little sign of rhinoceros beetle damage around there, so I took Wong, a field assistant, and proceeded in my U.N. Landrover to Kudat, a five-hour trip, on a twisty road, through forest as well as open country. At Kudat there were twenty 100-acre plots of coconuts owned by Chinese. In the absence of suitable breeding places rhinoceros beetle damage was minimal.

On our way back at Bandau we cut down a *Corypha* palm

Gan, mostly cleared of coconut palms to make an RAF staging post, must have been a haven for rhinoceros beetle shortly afterwards. There was plenty of it still in the few remaining coconuts, but nothing of much import. I visited a neighbouring island, Hittadu, where there were coconuts in the small RAF transmission concession. Otherwise the island was closed to Europeans. I procured 30 rhinoceros beetles to take back to Singapore for dissection. I had to be on the island five days and was able to buy duty-free Dewar's whisky for 14/6d and 'Lucky Strike' cigarettes for 1/2d.

After a five-hour flight back to Singapore I went to Kuala Lumpur where I did the dissections of the Maldive beetles; none had sex nematodes. I had left both species of beetles in cages together while I was at Gan.

Mt Kinabalu, from the rest house at Kota Belud, Sabah

— here called 'gebanga'. It had a widish trunk, the outer inch of which has fibres like iron. One could cut off only small chips with an axe, but once through that perimeter layer the fibres were fairly soft and easily cut. It took two hours work to bring it down. As it had completed its single terminal flower it was moribund and already contained 70 rhinoceros beetle larvae in the top five feet.

At the rest house at Kota Belud, where we spent the night, I got up at 3 am to see the Southern Cross just showing above the great mass of Mount Kinabalu — an impressive sight. On the way back to Tuaran, there were clumps of the sessile palm *Eugeissona*. It was the first I had seen of this palm. *O. gnu* attacked it and was also in coconuts thereabout. Rhinoceros beetle feeds on this palm as well. Thus there

was a good haul of beetles to dissect at Tuaran, but they all produced only expected results.

I spent Christmas with Tay and the Conways at Kundasang Guest House. This is about 5000 ft up on a hill facing the whole granite ridge of Kinabalu, the highest mountain in Malaysia (13,680 ft). From Jesselton, with Tay, I went to Tawau. Here were many deposits of sawdust, dead coconuts and other palms, a good site for future work. Owing to a delayed flight back we had to sit around in wet clothes, so I started the New Year with a pouring cold.

At Tuaran I tried to get rhinoceros beetle adults to make captive flights. This was an idea of Hoyt's. He thought perhaps some hypothetical tachinid might lay an egg on a soft part of the rhinoceros beetle anatomy exposed in flight. Knowing the attenuated population density of rhinoceros beetle it seemed like a chance in a million. With a leg tied to a fine nylon line, itself tied moveably to another, the beetles refused to take off. I had to abandon this experiment after at least attempting to comply with Hoyt's instructions.

I made a brief visit to Kuching, where a long hard walk with George Rothschild produced very few beetles. Thence I flew to Singapore to meet Noel, who once again was to assist me for the remainder of my time with the project.

We flew back to Jesselton, past the flaming oil wells of Brunei, and then on to Tuaran. Here we occupied the rest house and had the excellent Rusi to cook and wash for us. A.W. Allen, the Director of the Station, whose wife was on leave, made us a curry of the Malay style, with all the little side dishes that go with it — quite delicious.

Clumps of palms, 'Nibong' *(Oncospermum)* and

Eugeissona were both harbouring *O. gnu*, some with sex nematodes, but nothing to suggest diminished fecundity of the beetle.

At Tawau, we signed on two local young men: a Bugis called 'Shutim' and a Malay called 'Kassim'. Shutim would put his ear to a short dead coconut trunk and could hear if rhinoceros beetle grubs were present chewing at the dead wood. On one occasion I took them in a launch across a bay to examine stems of nipa. On the way back to Tawau the wind got up and the launch had to head into alarming waves. I was afraid we would founder, but somehow we kept going. I think the feeling that we had been through some daunting times together prompted these delightful young men to present us with a very realistic carving of two egrets in a courting position with beaks erect. We have prized this piece ever since.

Shutim listening to grubs of the red palm weevil, Tawau, Sabah

At the Quoin Hill cocoa research station, up-country from Tawau, we met the Director, Wyrlie-Birch, his wife and two children. We did collecting trips in our Landrover through tall dipterocarp forests, taking the children for the ride. En route to Semporna we had to board a rickety ferry. The kids enjoyed the rides and the sandwich lunches we had with them.

Among all the sawdust deposits near a mill I had a large cage made in which we released rhinoceros beetle adults. We tried captive flights again with fine nylon line fixed to a leg. *O. gnu* was a better performer than *O. rhinoceros* under such conditions, though neither of them enjoyed such games and were unwilling subjects.

At this time there was war between Malaysia and Indonesia and our scene of operations was near the frontier. We often met soldiers, but were submitted to no restraints — a contrast to the conditions I had had to undergo in the Moluccas in 1959.

At Wakuba, not far from Tawau, a dead *Corypha* palm trunk contained some dipteran larvae which appeared to be feeding on rhinoceros beetle grubs. Adults were bred and identified as a mydaid. This was the only predator on rhinoceros beetle we ever found. It would have taken a long time to rear them in any quantity and as Hoyt was wanting enemies of adult beetles it was not worth pursuing the possibilities of this larval predator.

After Sabah we were three days at Kuching. With Rothschild I went down the river and saw much rhinoceros beetle damage near a sawmill. *O. gnu* was not found. On from there to Singapore and thence to Kuala Lumpur to wind up our business with Sedky, Bickerstaff and Yunus.

Next was a flight from Singapore to Darwin. This town we found unpleasantly hot, the sun reflected from its whitish paving. On a visit to Bathurst Island we looked at two tall coconuts with a leaf or two looking as if they might have rhinoceros beetle cuts, but this really could not be confirmed. I was told that along the Liverpool River flowing into the Gulf of Carpentaria there were palms with fan leaves which seemed likely to be *Corypha*. I flew to Maningrida where there is an Aboriginal Mission. In a powered canoe trip up the river I saw the stand of *Corypha* palms, but none of the leaves had any rhinoceros beetle cuts, further evidence that rhinoceros beetle was not in Australia.

From Darwin we flew to East Timor, then in Portuguese control. At Dili, the capital, a lady in the tourist shop spoke a little English; but we got in touch with the Australian Consul, John Colquhoun-Denvers. When we told him our business he seemed interested to come and search for rhinoceros beetle with us in his car. There was no rhinoceros beetle damage thereabouts in *Corypha*. Another fan-leaf palm, *Borassus*, was abundant, the stems of them used as telegraph poles.

In an elevated situation, inland from Dili, I found a single beetle of *O. gnu* in an

Arengga palm — possibly the extreme easterly range of this species.

The agriculture people kindly lent us Edgar Sousa, an English-speaking local, who came with us to the eastern extremity of the island. There was rhinoceros beetle damage evident on coconut, *Corypha* and *Borassus*, leaving fascinating patterns of leaf cuts in these fan palms which would have made good fabric patterns. In dissections of adults there was the typical lack of sex nematodes.

Houses in the extreme east of Timor are built on piles to allow a passage of air beneath them in this area of high humidity.

From Timor we returned to Darwin, then on by air to Alice Springs, Adelaide and Sydney. Leaving Noel to stay with our Canberra family, I visited Cape York, Thursday Island and later Daru with Mick Catley. While at Cooktown I was told that there was a patch of *Corypha* near Laura, 50 miles inland. A light aircraft had brought two men on business from Cairns in the morning. As they didn't want their plane till the afternoon I was allowed to charter it to inspect the palms at Laura. Unfortunately, all the men there were away mustering cattle and the women knew nothing about the exact location of the palms I wanted to look at. After the pilot took off I could see the palms in the savanna below. I asked him to fly over them as low as he could. I looked with field-glasses, but could see no sign of rhinoceros beetle damage in the fronds, so concluded there was none.

At Daru, PNG, there were many dead coconuts, but no sago palms. We collected *O. centaurus* adults which contained only few sex nematodes. We crossed to the Pahuturi River on the mainland at Mabaduan. There were several different palms and pandans, some with *centaurus* larvae. A man at Guao said he could shoot 60 Birds of Paradise in a week and offered us a skin for three pounds. We told him he was breaking the law and we didn't want them. At Kadoro we met 15 men who were fugitives from Indonesians at Merauki.

While collecting beetles from coconuts I got an infected scratch. We crossed to Port Moresby where I took six aspirins and went to bed. Catley got a cable to say that Hoyt was ill and had to go to New Zealand, and he (Mick) was appointed Deputy Manager of the Project.

A last farewell to the Pacific

When I got back to Canberra I was treated with penicillin, that magic cure for ulcers. I was also treated for anaemia. After a few days at the hospital I got better and was able, with Noel, to fly to Noumea, headquarters of the SPC. Here we had to fix up the business side of my appointment. We then flew to Nandi where Mick told us of squabbles at Koronivia between Singh and Swain, the newly appointed entomologist. However, we had nothing to do with that matter and continued on to Apia. We stayed at Aggie's famous hotel. I wrote my report at Nafanua, after which, like in that Somerset Maugham story, we were prevented from leaving Samoa by days of heavy rain. Owing to that we missed out on an intended visit to Acapulco to see a little of Mexico. All we could do was to fly to Nassau, via Houston, New Orleans and Miami. It all felt so horribly sophisticated after our days of wanderings in 'the Bush'.

So ended the last of our post-war Pacific Island projects extending over the ten years to 1966. What a memorable time it had all been.

PUBLICATIONS
R.W. PAINE

Paine, R.W. 1932. Entomological report to the Coconut Committee. Fiji Department of Agriculture, Divisional Report. Annual Bulletin (1931), 1–6

Paine, R.W. 1932. Entomological notes. Fiji Agricultural Journal, 5, 3.

Paine, R.W. 1934. The control of Koster's curse *(Clidemia hirta)* on Taveuni. Fiji Agricultural Journal, 7, 10–21.

Paine, R.W. 1935. The control of the coconut spike moth *(Tirathaba trichogramma* Meyr.) in Fiji. Fiji Department of Agriculture Bulletin, 18, 30 p.

Paine, R.W. 1960. Observations on the banana moth *(Nacoleia octasema,* Meyr.) in Indonesia, and on the introduction from Flores to Java of its parasite *Chelonus striatigena,* Cam. Pemberitaan Balai Besar Penjelidikan Pertanian, 160, 33 p.

Paine, R.W. 1961a. Search for an insect parasite on the Northern Bangalow Palm. North Queensland Naturalist, 29(130), 6–7.

Paine, R.W. 1961. Observations on the banana scab moth in the Territory of Papua and New Guinea. Papua and New Guinea Agricultural Journal, 14, 45–47.

Paine, R.W. 1964. The banana scab moth *Nacoleia octasema* (Meyrick); its distribution, ecology, and control. South Pacific Commission Technical Paper, 145, 70 p.

Paine, R.W. 1966. Report on a search for parasites and predators of *Oryctes* and related dynastids in the S.E. Asia region. Report to U.N.D. (S.F.)/SPC Project for Research on the Control of Coconut Rhinoceros Beetle. Typescript, Unpublished, 64 p.

Paine, R.W. 1968. Investigations for the biological control in Fiji of the coconut stick-insect *Graeffea crouanii* (Le Guillou). Bulletin of Entomological Research, 57, 567–602.

Tothill, J.D., Taylor, T.H.C. and Paine, R.W. 1930. The coconut moth in Fiji. A history of its control by means of parasites. Imperial Bureau of Entomology, Imperial Agricultural Bureaux, 269 p.